REFRAMING CLIMATE CHANGE

"Change the system, not the climate" is a common slogan of climate change activists. Yet when this idea comes into the academic and policy realm, it is easy to see how climate change discourse frequently asks the wrong questions. *Reframing Climate Change* encourages social scientists, policy-makers, and graduate students to critically consider how climate change is framed in scientific, social, and political spheres. It proposes ecological geopolitics as a framework for understanding the extent to which climate change is a meaningful analytical focus, as well as the ways in which it can be detrimental, detracting attention from more productive lines of thought, research, and action.

The volume draws on multiple perspectives and disciplines to cover a broad scope of climate change. Chapter topics range from climate science and security to climate justice and literacy. Although these familiar concepts are widely used by scholars and policy-makers, they are discussed here as frequently problematic when used as lenses through which to study climate change. Beyond merely reviewing current trends within these different approaches to climate change, the collection offers a thoughtful assessment of these approaches with an eye towards an overarching reconsideration of the current understanding of our relationship to climate change.

Reframing Climate Change is an essential resource for students, policy-makers, and anyone interested in understanding more about this important topic. Who decides what the priorities are? Who benefits from these priorities, and what kinds of systems or actions are justified or hindered? The key contribution of the book is the outlining of ecological geopolitics as a different way of understanding human–environment relationships including and beyond climate change issues.

Shannon O'Lear is a Professor at the University of Kansas, USA, where she has a joint appointment in the Departments of Geography and Environmental Studies. She is the author of *Environmental Politics: Scale and Power* (2010). She has published widely on energy and natural resources, environmental security, and critical geopolitics of the environment.

Simon Dalby is CIGI Chair in the Political Economy of Climate Change at the Balsillie School of International Affairs and Professor of Geography and Environmental Studies at Wilfrid Laurier University, Waterloo, Canada. His previous books include *Creating the Second Cold War* (1990), *Environmental Security* (2002), and *Security and Environmental Change* (2009).

"Climate change means different things to different people in different places. It does not have one cause and one solution, but many causes and many solutions. In *Reframing Climate Change*, O'Lear and Dalby have brought together an impressive group of political geographers and scientists who undermine the conventional singular narrative of climate change – 'the plan' as some have dubbed it – before helpfully opening up different ways of framing what is at stake. It is only with such a pluralist account of climate change that the business of politics can get done: to expose, argue over and decide between different visions people have of how the world should be."

Mike Hulme, Professor of Climate and Culture in the
Department of Geography, King's College London, UK

"This book is essential reading for those who see or sense that climate change is more than an environmental issue. The authors argue that it is critical to challenge conventional framings of both problems and solutions by bringing in politics, power and new perspectives. *Reframing Climate Change* unravels some key assumptions that have the potential to transform approaches to security in the Anthropocene."

Karen O'Brien, Professor in the Department of Sociology
and Human Geography, University of Oslo, Norway

REFRAMING CLIMATE CHANGE

Constructing ecological geopolitics

Edited by Shannon O'Lear and Simon Dalby

LONDON AND NEW YORK

First published 2016
by Routledge
2 Park Square, Milton Park, Abingdon, Oxon OX14 4RN

and by Routledge
711 Third Avenue, New York, NY 10017

Routledge is an imprint of the Taylor & Francis Group, an informa business

© 2016 Shannon O'Lear and Simon Dalby

The right of the editors to be identified as the authors of the editorial material, and of the authors for their individual chapters, has been asserted in accordance with sections 77 and 78 of the Copyright, Designs and Patents Act 1988.

All rights reserved. No part of this book may be reprinted or reproduced or utilized in any form or by any electronic, mechanical, or other means, now known or hereafter invented, including photocopying and recording, or in any information storage or retrieval system, without permission in writing from the publishers.

Trademark notice: Product or corporate names may be trademarks or registered trademarks, and are used only for identification and explanation without intent to infringe.

British Library Cataloguing in Publication Data
A catalogue record for this book is available from the British Library

Library of Congress Cataloging in Publication Data
Reframing climate change : constructing ecological geopolitics / edited by Shannon O'Lear and Simon Dalby.
    pages cm
  Includes bibliographical references and index.
  1. Climatic changes—Political aspects. 2. Climatology—Political aspects.
  3. Geopolitics. I. O'Lear, Shannon. II. Dalby, Simon.
  QC903.R429 2015
  363.738'74—dc23   2015013423

ISBN: 978-1-138-79436-8 (hbk)
ISBN: 978-1-138-79437-5 (pbk)
ISBN: 978-1-315-75926-5 (ebk)

Typeset in Bembo
by Keystroke, Station Road, Codsall, Wolverhampton
Printed by Ashford Colour Press Ltd.

CONTENTS

List of figures   ix
List of contributors   xi
Preface   xv

1 Reframing the climate change discussion   1
   Shannon O'Lear and Simon Dalby

2 Postmodern interpretations   14
   Leigh Glover

3 The climate of communication: from detection to danger   31
   Chris Russill

4 Disconnecting climate change from conflict: a
   methodological proposal   52
   Emily Meierding

5 Climate justice: climate change, resource conflicts,
   and social justice   67
   Paul Routledge

6 Climate change and the insecurity frame   83
   Simon Dalby

7 Geopolitics and climate science: the case of the missing
   embodied carbon   100
   Shannon O'Lear

8	Technology and politics in the Anthropocene: visions of "solar radiation management" *Thilo Wiertz*	116
9	Biofuels: climate solution or environmental pariah? *James Smith and Shaun Ruysenaar*	132
10	Novel framings create new, unexpected allies for climate activism *Andrew Szasz*	150
11	Catastrophe insurance and the biopolitics of climate change adaptation *Kevin Grove*	171
12	Resisting the climate security discourse: restoring "the political" in climate change politics *Angela Oels*	188
13	Towards ecological geopolitics: climate change reframed *Simon Dalby and Shannon O'Lear*	203

Index   *217*

FIGURES

3.1 Poster for *An Inconvenient Truth* (2006) 32
3.2 Reasons for concern over climate change pegged to increases in global mean surface temperature 34
9.1 The governance–government continuum and instruments of governance 139

CONTRIBUTORS

Simon Dalby is CIGI Chair in the Political Economy of Climate Change at the Balsillie School of International Affairs and Professor of Geography and Environmental Studies at Wilfrid Laurier University, Waterloo, Canada. He is the author of *Creating the Second Cold War* (1990), *Environmental Security* (2002), and *Security and Environmental Change* (2009).

Leigh Glover is an academic and researcher. He is a former Director of the Australasian Centre for the Governance and Management of Urban Transport (GAMUT) at the University of Melbourne, Australia. He has authored several chapters, papers, and reports on climate change and the book *Postmodern Climate Change* (2006).

Kevin Grove is a Lecturer in Human Geography in the Department of Geography and Earth Sciences at Aberystwyth University, UK. His research focuses on the biopolitics of disaster management, and especially on catastrophe insurance and community-based disaster resilience programming in the Caribbean. He has published in geographic and interdisciplinary journals, including *Annals of the Association of American Geographers*, *Security Dialogue*, *Environment and Planning D: Society and Space*, and *Resilience*.

Emily Meierding is a Visiting Fellow at the Centre for International Environmental Studies at the Graduate Institute of International and Development Studies, Geneva, Switzerland. Her research examines the relationships between traditional energy resources, climate change, and conflict. She has published in the *International Studies Review*.

Shannon O'Lear is a Professor in both the Geography Department and Environmental Studies Program at the University of Kansas, USA. She is the author of *Environmental Politics: Scale and Power* (2010). She has published on energy, resource, and identity issues in the South Caucasus with a focus on Azerbaijan, territorial conflict, genocide, borders, environmental security, and critical geopolitics of the environment.

Angela Oels is a Visiting Professor in the Department of Political Science at the Lund University Centre on Sustainability Studies (LUCSUS), Sweden. Her current research focuses on the construction of climate change as a security issue in international political and scientific discourses, with a particular emphasis on climate change-induced migration. She has published in journals of environmental politics, geography, political science, and international relations, including *Security Dialogue*, *GEOFORUM*, and the *Journal of Environmental Policy and Planning*.

Paul Routledge is a Professor in Social and Urban Change in the School of Geography at the University of Leeds, UK. His research interests include critical geopolitics, climate change, social justice, civil society, the environment, and social movements, with a focus on the Global South, particularly South Asia and Southeast Asia. He is the author of *Terrains of Resistance: Nonviolent Social Movements and the Contestation of Place in India* (1993), co-author (with A. Cumbers) of *Global Justice Networks: Geographies of Transnational Solidarity* (2009), co-editor (with J. Sharp, C. Philo, and R. Paddison) of *Entanglements of Power: Geographies of Domination/Resistance* (2000), and co-editor (with G.O. Tuathail and S. Dalby) of *The Geopolitics Reader* (1998, 2006).

Chris Russill is an Associate Professor in the School of Journalism and Communication at Carleton University, Ottawa, Canada. He is the editor of "Earth-Observing Media", a special issue of the *Canadian Journal of Communication*, and author of many papers on the imaging and communication of global environmental crises, including hurricanes, ozone holes, and climate change.

Shaun Ruysenaar completed his Ph.D. at the University of Edinburgh, UK, and is currently Chief Operations Officer at JFKS Consulting and an Associate Researcher for the School of African Studies at the University of Edinburgh. His thesis focused on the social and political dynamics of policy-making processes around biofuels in South Africa. He has published on food security and biofuels development in South African and international journals and has broad interests in governance, science, and technology and development studies.

James Smith holds a personal chair in African and Development Studies and is Vice-Principal International at the University of Edinburgh, UK, as well as holding posts at the University of Johannesburg, South Africa, and the Open University, UK. His research focuses on the interrelationships and tensions between science, innovation, and international development, and he has published extensively on topics including agricultural biotechnologies, vaccine development, zoonotic disease control, and biofuels. He is the author of *Biofuels and the Globalization of Risk* (2010).

Andrew Szasz received his BA from Harvard College, USA, in 1969 and his Ph.D. in Sociology from the University of Wisconsin, Madison, USA, in 1982. He is currently Professor and Chair of the Department of Environmental Studies at the University of California, Santa Cruz, USA. He has written books and articles on environmental regulation, grassroots toxics movements, green consuming, environmental justice, and, most recently, the sociology of climate change. He teaches courses on environmental justice, the sociology of climate change, the history of the American environmental movement, and sociological theory.

Thilo Wiertz is an Associate Researcher at the Institute for Advanced Sustainability Studies (IASS) in Potsdam, Germany. He is part of an interdisciplinary research group on geoengineering. His research interests lie within the fields of political geography, society–nature relations, and science and technology studies.

PREFACE

Not so many years ago, there were only a few political geographers working on topics related to the environment. Geographers have long been engaged in questions of human–environment relationships, but it is only recently that we have seen an upwelling of interest in scholarly inquiry into dynamics of power and space in myriad forms and with tangible implications. This has taken place in recent years as scholars with training in other disciplines have also begun to think seriously about environmental change and its consequences.

In the last decade, with the realization of just how dramatically humans are changing things, especially in terms of climate, the need to address these big questions has become compelling across the disciplines. Thus, while this volume is edited by two political geographers, it reaches across other scholarly fields to ask probing questions about both how we frame climate change in contemporary politics and policy discussions, and how we can now examine these from a multiplicity of scholarly perspectives.

This volume reflects the efforts of each of the contributing authors to stand back and think critically about a sprawling and complex issue that touches us all. We are grateful for the creative work of these scholars, who have taken unconventional approaches to their topics so that we may all benefit from fresh perspectives and analyses that question – rather than accept – the ways in which current modes of power in decision-making, communication, finance, governance, and policy-making can serve as obstacles to more just and practical responses to our current context.

We do not face an imminent time of climate change; we are living within it. Atmospheric alteration is a symptom of our way of life and reflects political, economic, and social processes that have become accepted as the norm. Our hope is that the chapters in this book open pathways for further exploration within and beyond the field of political geography that will provide a foothold from which we can act in meaningful ways to tackle the challenges of climate change.

The task of clear writing, Steven Pinker (2014: 38) suggests, is like "the celebrity chef in the immaculate television kitchen who pulls a perfect soufflé out of the oven in the show's final minute"; "the messy work has been done beforehand and behind the scenes." We are most grateful for the editorial grace that Eve Nimmo has brought to this effort. It is through her efficiency and thoroughness in editing all of the chapters that this creative effort has been brought into tidy and cohesive shape. Anyone who has attempted to generate an index for a book will understand our appreciation for her energy and attention to detail despite the continental distance that separates us. Thank you, Eve, and to the Balsillie School of International Affairs for a grant in aid of publication for making it possible to involve her in this project.

Thanks also go to our students and colleagues, who ask questions and challenge us to explain the "So what?" of our research. It is through these interactions in classrooms, along corridors, and at conferences that we are often inspired to look at something in a different way, clarify our own thinking, or seek a new way to ask and respond to questions that might otherwise go unasked.

More specifically, Leigh Glover would like to thank Dr Mikael Granberg at Karlstad University, Sweden, for his advice. Angela Oels thanks Eve Nimmo for help with language. James Smith and Shaun Ruysenaar would like to thank the UK Economic and Social Research Council, the UK Department for International Development, and the Commonwealth Scholarship Commission for supporting elements of the research that inform their chapter. Chris Russill acknowledges conversations with Shane Gunster, Mike Hulme, and Naomi Oreskes as particularly helpful without implicating them in any of the claims made in his chapter – each of his interactions with them shaped his thinking in important ways. He also thanks his colleagues in the Carleton Working Group on Climate Change, who have offered continual insight and inspiration, and the Balsillie School of International Affairs, which provided a stimulating environment in which to complete his chapter. Thilo Wiertz would like to thank the Marsilius Kolleg at Heidelberg University and the Institute for Advanced Sustainability Studies in Potsdam for supporting his research.

Working with Routledge has been a pleasure thanks to support and assistance from our editor, Andrew Mould, and editorial assistant, Sarah Gilkes. Thanks also to photographer Julie Dermansky, who helped to visualize our message on the book's cover.

Shannon O'Lear and Simon Dalby

Reference

Pinker, S. (2014) *The Sense of Style: The Thinking Person's Guide to Writing in the 21st Century*, New York: Viking Press.

1
REFRAMING THE CLIMATE CHANGE DISCUSSION

Shannon O'Lear and Simon Dalby

Introduction

On 6 July 2013 a train carrying crude oil derailed in the town of Lac Megantic in Quebec, causing explosions and fires that killed 47 people and destroyed the centre of the town. While petroleum does not usually explode in such a fashion, it turned out that this oil came from the Bakken oil field in North Dakota. At the time, its exploitation was creating an economic boom and causing a rapid demand for workers, housing, and supporting service industries. Because pipelines were not available for the highly flammable Bakken crude, it was being shipped by train. Oil extraction was happening so quickly that it was not even profitable to capture the natural gas, so it was burned off as a waste product. The impacts were visible from space; on colour satellite images taken at night, the natural gas flares near the oil rigs appeared as red lights. Some of these gas flare areas were so big that they competed in size with the lit-up twin cities of Minneapolis and St Paul. The fires in Quebec were also briefly visible from space in July 2013. The amount of natural gas being flared off daily was enough to heat three-quarters of a million homes.

Elsewhere in North America at the same time, many corporations and cities were attempting to trim their carbon footprints to promote the notion of sustainability; and consumers were encouraged to pay a fee that would somehow offset the carbon generated by a plane flight or to otherwise be "green" in their consumption. Even the US military was looking for ways to reduce its energy consumption and costs on bases and posts across the country. Reducing consumption and using fuel efficiently was supposedly the order of the day. At least it was for some of North American society, but apparently not for the petroleum industry in the midst of an economic boom fuelled by high prices. Eighteen months later, as world oil prices plummeted, investments in new oil and gas production slowed. Financial incentives for consumers to reduce fuel consumption and hence slow climate change were substantially reduced, too.

These trends would appear to be working in different directions. How are we to make sense of the various links between human activity and climate change? Larger questions raised by such issues are both profound and troubling. They are profound because we are gradually realizing that humanity is changing the fundamental conditions of its existence. They are troubling because the current ways of thinking seem both inadequate to deal with the new circumstances and trapped in political, social, and economic modes that now seem singularly inappropriate. Climate change apparently requires acting in ways that might prevent at least the most obvious damage and harm to humans and other species in the immediate future, but our modes of thought and political structures work to perpetuate precisely the economic and political arrangements that are causing the potentially dangerous changes to the climate system.

As we try to understand how we have come to this pass and what to do about it, scholars and academic researchers in many of the sciences ought, surely, to have answers and insights as to how to proceed. Yet, even scholarly and scientific approaches to understanding and assessing climate change seem to have become "trapped" by methodological norms, disciplinary boundaries, and a narrow focus on small parts of the problem at the expense of larger synthetic understandings. A quarter of a century of huge and increasingly alarming synthesis reports on the scientific knowledge of climate change by the Intergovernmental Panel on Climate Change (IPCC) has produced no effective curbs on the production of greenhouse gases (GHG) in the global economy. Simply presenting the facts and assuming that politicians will behave sensibly in response apparently has not worked. Moral exhortation does not seem to have been any more effective.

The fossil fuel companies, the producers of so many of the GHG products that power the global economy and most urban dwellers' lives, are focused on continuing their production and their profits, not on how to decarbonize economies rapidly. Competition and consumption are key to capitalist growth and literally drive the global economy. While there are innovative market mechanisms that might move the fuel mix away from reliance on coal and petroleum in particular, current energy transitions seem far too slow to begin the shift to something that might meaningfully be called "ecologically sustainable."

Addressing climate change through international agreements is also a lagging effort. Promises of some kind of treaty that will be binding on all states have been repeatedly deferred. Only in 2014 did China agree to firm caps on its emissions levels in future decades. This came 20 years after the adoption of the United Nations Framework Convention on Climate Change (UNFCCC). Even if ambitious attempts to formulate a climate treaty in Paris late in 2015 come to fruition, most plans suggest that it will enter into effect only in 2020. Many developing states are more worried that such agreements might hamper their economic development than they are about climate changes that they have done little to cause and the course of which they can do little to alter. The more powerful states are, it seems, still more concerned with the traditional questions of rivalries and influence in global politics than they are with what kind of planet we are now collectively making for future generations.

Often, academic approaches to climate change apply methods suited to other problems or build on discourses that now seem to ask the wrong questions. The agendas and findings of the physical sciences are often politicized in ways that favour the status quo (Sarewitz 2004) rather than motivate necessary social and economic adjustment. The social sciences tend to be concerned with practical problems of routine public administration, social stability, and economic growth, rather than matters of how societies can adapt to climate change, much less how we might make sustainable social arrangements that do not rely on fossil fuels for power.

Given the neoliberal times in which we live, where markets are the preferred modes of governance, citizenship is frequently reduced to consumption, and success is assessed in short-term financial measures. Thinking that we can "shop our way to safety" flies in the face of more serious attempts to grapple with the changing ecological circumstances of our times (Szasz 2007). But grapple with these changing circumstances we must, as students, scholars, and citizens of various parts of an increasingly interconnected world.

Climate change and its implications have no obvious analogy in human affairs. Despite this, we mostly continue to apply governance solutions, attempt to securitize it, and frame it within Cold War geopolitical understandings of security, economic calculations of risk, or traditional formulations of environmental governance, such as pollution control. Political analyses often perpetuate the focus on rivalries and power politics by simply adding climate change into the traditional investigations of international regimes, rivalries, and treaty-making rather than thinking fundamentally about how ecological change is remaking the conditions for states, economies, and societies (Hommel and Murphy 2013).

Only rarely, still, is it acknowledged that climate change is an evolving social and physical phenomenon that cannot be addressed by "a" negotiated solution (Hulme 2009). It is precisely the growing cogency of analyses that understand climate change as part of a larger series of physical and social transformations coupled with the recognition that there is no single "solution" to the "problem" of climate change that gives rise to this book.

Welcome to the Anthropocene!

Climate change is but one part of the larger transformation of the planet set in motion by human activities (Steffen et al. 2011). We have already crossed some key boundaries in terms of how the planet works that suggest that humanity is starting to live in circumstances that are now rather different from the ecological conditions that gave rise to human civilization. In the words of the earth system scientists who study the planet as a whole, we are living in a new geological epoch frequently called the Anthropocene (Crutzen 2002). This geological era is marked by significant changes not only to the chemical composition of the atmosphere, but to systems of heat exchange, acidification of oceans, levels of pollution, and alterations to landforms, water flows, nutrient cycles, and flora and fauna dynamics across the planet.

The Industrial Revolution fostered a carbon fuel-powered economy that thrived literally on our ability to "turn rocks into air" (Dalby 2009: 71). As industrial practices diffused and the rate and reach of globalization increased, so did the human impact on the planet. Viewed in this way, we can no longer understand ourselves in modern terms as using our technologies to alter an environment external to ourselves. Instead, we now have to understand ourselves as part of an earth that our actions are rapidly changing. Such insights require rethinking many things, not least our assumptions about a given environmental context within which humanity functions.

This book offers a series of critical examinations of how we understand – or frame – climate change as a scientific, social, and political matter. Such critical perspectives examine commonly held understandings and assumptions in order to assess how and why certain views are promoted and why selected information is prioritized or marginalized. Who benefits from these priorities, and what kinds of systems or actions are justified or hindered? These are underlying questions directed towards climate change throughout the book. More specifically, the project emphasizes spatialities and power dynamics associated with climate change discourse, research, and policy. As with earlier work on resource conflicts, it is important to try to ask the right questions, rather than take the given policy and administrative frameworks for granted as the premises from which to do scholarly work (O'Lear and Gray 2006). While this book may not ask all of the right questions, clearly it is necessary to question rather than simply accept the dominant framings of climate change.

The premise for this book is quite simple: we, as scholars in a number of disciplines, need to reassess how we study, interpret, and respond to climate change so that we can develop more appropriate ways of thinking about society–environment relationships. Climate change is not merely an issue to be measured and understood by scientific means. It is also a deep reflection of political and economic disparities as well as the current spatial connections and disconnections of human activity. A focus on climate change without an understanding of the geopolitical and economic dynamics that have generated unprecedented environmental change – as well as the scientific systems and practices that are used to interpret climate change – cannot fully inform appropriate policy, scholarship, and citizenship. Understanding who we are as citizens, consumers, and scholars, and how existing social, economic, and political structures shape our identities and practices, and how they might be changed, is also key to reframing climate change as a matter of remaking the planet and ourselves simultaneously.

A key contribution of this book is our proposal to rethink these kinds of things under the rubric of "ecological geopolitics." This approach is suggested to aid understanding of the extent to which climate change is a meaningful focus for more productive lines of thought, research, and action. A focus on ecological geopolitics emphasizes that humanity is an increasingly influential part of the planetary system and that decisions about what gets made, how landscapes are reshaped, how buildings, infrastructure, and commodities are produced, and how species, hydrologies, and chemicals are reassembled in new configurations are creating the planetary context of our times. Ecological geopolitics offers possibilities for different ways of understanding human–environment relationships including and beyond climate change issues. This focus shifts attention to power dynamics that serve to reinforce uneven distributions of human and ecological well-being.

Such thinking urgently needs to be undertaken now, given the failure of conventional social and physical science to tackle the profound questions posed by the Anthropocene. Whenever a major storm disrupts human activities, attention focuses on how to frame questions about the relationships between climate change and extreme events (Trenberth 2012). Sustained attention is needed, however, and the authors of the following chapters contribute to the process of rethinking and reframing climate change that now seems so necessary.

Framing matters

Much discussion of discourse and representation in the social sciences now draws on formulations of framing to investigate how things are constituted in social life (Goffman 1974). We draw on these insights loosely in the pages that follow to emphasize that what is considered climate change is not so obvious. As with the metaphor of frames, what is inside the frame of a picture is a choice. What is focused on, included as part of the image, is distinguished from what is cropped, excluded, left outside the frame. More than that, as anyone who has ever had a piece of artwork "framed" understands, how the frame is constructed – the texture and colour of the matting and the frame itself, as well as the type of glass – emphasizes certain aspects of the image and leads the viewer's eye to see things in particular ways. While the metaphor can probably be stretched even further, the point here is to emphasize how climate becomes a matter of scientific and social knowledge; how it is talked about, discussed in the media and especially by academics, is a matter of selectivity. Climate change is a social construction (Onuf 2007). It is not a given entity, but something known and disputed in part by how discourses frame the subject matter (Strauss and Orlove 2003).

In policing, law, and discussions of crime, framing has a rather different connotation. Being framed is a matter of being portrayed as guilty as a result of deliberate planting of false evidence, fictitious testimony, or deliberate falsehoods to shift the blame for a crime onto an innocent party. Constructing a prosecution case, which is effectively an argument concerning responsibility and hence guilt, on the basis of manufactured evidence, a "frame" is widely understood as an unacceptable practice, a matter of obstructing justice. Matters are less clear when investigators misinterpret evidence, make mistakes, or jump to conclusions without access to all of the necessary information to develop an accurate understanding of events and responsibilities. Extending the metaphor of framing in these ways suggests that social practices – how things are constituted as entities and put into relationships one with another – are key processes in the social production of knowledge, which in turn is related to how identities and social roles are constructed and matters of responsibility assigned. Such frames may intentionally or unintentionally shape understanding. In discussing climate change, even assuming that this term is an appropriate frame for contemporary circumstances, or an entirely objective condition, is an important consideration.

Our purpose in phrasing matters in this way is not to suggest, as the so-called climate deniers frequently do, that climate science is false, or, in US Senator Jim Inhofe's infamous phrase, "a hoax." We emphasize framing because how objects are constituted, and the power dynamics that result from those formulations, is an unavoidable part of the discussion of climate change. Additionally, the role of science, criticism, and knowledge in political dispute is part of what has to be tackled in any attempt to think critically about what is now called "climate change." The framing metaphor also emphasizes that how stories about climate change are told to spell out who did what, and how responsibility can thus be assigned and with what social consequences, is crucial to discussions of justice, governance, and security.

How security, justice, governance, and economy have been framed in the past matters greatly in understanding how current discourses of climate change play out. The assumptions about the appropriate role of government, how threats to social order are formulated, who is responsible for environmental change, and how it endangers particular peoples and social orders in specific places are all parts of the discussion. As Hulme (2009) makes very clear, disagreements about climate change are not just matters of dispute over scientific fact, but are deeply tied into cultural, political, and religious assumptions about both the larger conditions of humanity's existence and moral framings of who should act, how, and why in contemporary social arrangements. As the authors aim to tease out in the chapters that follow, these matters run through all aspects of discussions of climate that are unavoidably matters of interpretations, conflicts, claims about justice, and scientific and technological concerns as well as governance broadly construed.

Environmental security?

Climate change matters, so the conventional international framing has it, because it is potentially very dangerous to humanity. In the terms used in the UNFCCC, which was ratified by nearly all UN member states by 1994, it is important to prevent dangerous human interference with the planet's climate system. While that statement was widely agreed upon more than two decades ago, quite what constitutes dangerous interference is not so easy to define. The suggestion that the planet must not heat more than an average of two degrees Celsius over pre-industrial levels was used repeatedly in the 1990s, and was still widely accepted by states at the Copenhagen Climate Conference in late 2009.

While rapid cuts to emissions of GHGs seem essential to this task, numerous scenarios are in play concerning how this might be accomplished and with what consequences should these changes be delayed. Should the atmosphere be stabilized at 450 parts per million (ppm) of carbon dioxide, or perhaps at something much closer to the pre-industrial level of 280 ppm? In 2015, atmospheric concentrations were close to 400 ppm, up from 350 ppm when the Cold War ended. Even stabilizing at 450 ppm will need rapid changes to energy use, starting *now*. Recent authoritative analyses sponsored by the World Bank (2014) suggest that, unless things change dramatically, global temperatures will increase well beyond the two-degree-Celsius limit and humanity will be facing a dramatically altered world a few decades hence.

The principle that there are common but differentiated responsibilities in dealing with climate change has also been widely accepted, but who should act, and how, is not so clear. We understand climate change to be a global problem. After all, the atmosphere is singular; there are no multiple airs in conventional understandings of either meteorology or climate change. Climate models treat all carbon dioxide as the same, regardless of where it is, but this practice neglects political economic disparities underlying the patterns of carbon dioxide emission. Justice advocates and leaders of economically developing states make it clear that they do not accept that carbon dioxide emitted by poor farmers trying to feed their families in the Global South should be equated with luxury emissions by rich Northerners playing in their gasoline-powered recreational vehicles (Agarwal and Narain 1991, 1998).

However these matters are represented, they are not clear or obvious in terms of what needs to be done by whom, where, and when. But surely the overarching objective of security makes such things a priority. After all, security is about protecting people from dangers and setting in motion policies by states to deal with threats to social stability. Here, too, the legacy of past practices, not to mention incompatible definitions of security, is not very helpful. However, tackling the legacy of past formulations – and the implicit assumptions about who should act, and how, in the face of dangers – is a useful way to begin to rethink the framing practices that shape understandings of climate change. While environmental matters have periodically been understood in terms of their threats to various forms of security, in the past decade climate change has emerged as the key theme in discussions of high-priority concerns for many states in the international system (Webersik 2010). In short, climate change is increasingly a matter understood in terms of dangers requiring state-level policy actions in terms of security.

During the Cold War and its immediate aftermath, security was primarily understood by US planners in terms of preparing to deter or defeat military threats from other states, most obviously the Soviet Union. Other minor issues of policing and coercion among small states that were important suppliers of fuel and materials for the global economy were not to be ignored, but they were of lesser importance than preparing to fight other large state militaries. At least they were until the morning of 11 September 2001, when al Qaeda's suicide flyers spectacularly upset the geopolitical reasoning of much of the security establishment and began a rethinking in terms of "global" warfare and a greatly expanded range of non-traditional threats. A global "war on terror" followed, one that has militarized numerous facets of security and shaped governance structures in ways that emphasize state-centred security as the default policy mode. In subsequent years this mode of security thinking has portrayed climate change as a threat multiplier (see O'Lear et al. 2013) or a conflict multiplier in which instabilities in peripheral places might feed conflict or terrorism (Chalecki 2013).

The new security agendas in the twenty-first century do not fit the state military rivalry models of traditional international relations. In the case of climate change, a basic problem is that it is a symptom of the success of modern states even if some of its causes, such as deforestation, are indirect results of the global political economy. This is the case because the economic systems of modern capitalism are based largely on the use of cheap fossil fuels: first coal, subsequently petroleum, and most recently natural gas, the use of which is elevating atmospheric levels of GHGs that are driving climate change. States are part of the cause of climate change, and security framings tend to focus on states as the priority or solution. Here we might consider a thought attributed to Albert Einstein: we cannot solve problems by using the same kind of thinking we used when we created them. What thinking, then, should we use?

Ecological geopolitics

The sheer scale of the transformations that are now under way is slowly dawning on us, and the potential disruptions to current modes of life now qualify clearly as matters of security understood in the terms of collective vulnerabilities related directly to the dynamics of the global economy (Stiglitz and Kaldor 2013). Until recently, few people beyond the climate change scientific community had appreciated that humanity is changing some of the basic parameters of the planetary system at rates that are unprecedented in geological history. Indeed, we are entering a planetary phase shift of unknown duration and outcome (Barnosky et al. 2012). These are not matters of slow, linear transformations that can be left to future generations to resolve (Mayer 2012). It is becoming clear that climate change is happening now.

In Stiglitz and Kaldor's (2013) terms the quest for security in these circumstances is better tackled by cooperation rather than unilateral measures. However, it is far from clear that existing state functionaries can change either their geopolitical thinking or their economic policies enough so that they work cooperatively on climate change in time to prevent dramatic transformations of the biosphere. This political problem is an altogether larger set of issues than the traditional themes of global environmental governance that deal with such matters as the trade in endangered species, ozone depletion, toxic substances, and food safety concerns. Climate change raises profound questions of political economy and poses questions of how states can be transformed to respond and adapt to climate change whether as a security matter or as defined in some other way.

Fundamentally, globalization, which is both a matter of rapidly integrating economic and political activities, is also a matter of ecological transformation (Dalby 2009). While the expansion of European power in the age of empires was about inter-imperial rivalry and conflicts over access to colonies, resources, and markets, that understanding of geopolitics is now dangerously out of date. As state elites continue to wrestle with how to increase their power in a competitive system, the context in which that competition plays out is being transformed by economic and technological innovations that are shifting the priorities of states (Terhalle and Depledge 2013). While military rivalries continue – and so long as they do, major warfare between large states cannot be ruled out as a possible outcome – most of the competition now is about trade, the terms of trade, and arcane rules of financial and technical designations of items and processes in the flows of globalization. How climate change plays out in these arenas is only recently coming into focus. The creation of carbon markets, access to carbon offsets, and dominance in the new energy technologies, and how the rules for trading these technologies are written, are increasingly aspects of geopolitical calculation (Wang et al. 2012). How fears of shared dangers about climate change will shape these fora, or not, in coming decades is a key matter of international relations.

The implications of all of this are profound for the geographical assumptions built into how governance is now understood, the role of state security agencies, and the strategies that state policy-makers adopt. It is clear that if the planet is not to experience dramatic changes in the coming decades, security in the sense of preventing rapid, painful disturbances to social life will have to address consumption in the North first and elsewhere soon afterwards. Rapidly changing how energy is used and reducing the use of fossil fuels are key. This is not a matter of protecting a given environment; it is a matter of deciding what kind of planet will be made for future generations. It is a matter of production much more than of pollution, of decisions about what is built and how rather than of parks, nature reserves, and species loss prevention. The threats to humanity do not originate from distant, hostile forces; they come, in large part, as a result of the routine everyday lives of suburban, automobile-driving consumers.

Reframing security in these terms requires some dramatic rethinking of states, economies, threats, and politics. Geopolitics is now more a matter of shaping the conditions for future human life than of struggling over a finite amount of resources in games of great power prestige (Dalby 2014). While much of modern life and its economic understandings of reality assume that scarcity is the given human condition and struggles are an inevitable consequence, climate change makes it very clear that insofar as it is a "problem," it is caused by too much fossil fuel use, not too little. The massive production systems of the global economy are transforming terrestrial and oceanic life very rapidly. This is a matter of ecological transformation, not a matter of struggles over scarce resources, despite the fact that discussions over fuel supplies, water, and most recently rare earth minerals are reprised in the classic tropes of resource scarcity.

The social sciences have been slow to realize the geological scale of the transformation we are living in, and the physical sciences have not translated into a vision for meaningful social, economic, and political adjustment to our ecological circumstances (Vinthagen 2013). Our use of the term "ecological geopolitics" in the subtitle of this volume focuses attention on this new context. Starting the book with a reframing of geopolitics in terms of ecology is suggestive of the rethinking that numerous disciplines are beginning to undertake. These intellectual efforts are driven in part by climate change but also by intellectual and practical changes in how things are studied, understood, and communicated now that the simple dichotomies of modernity are no longer accepted as the starting point for either humanities or science (Braun and Whatmore 2010). We are seeing the results of a worldview that sets humans apart from other species, but in neglecting our inherent connections to the world around us, we inflict damage on ourselves as well. Ontological categories – such as humans as distinct from "the" environment – in both the social and physical sciences are in flux, and it is in this intellectual context that reframing climate change, whether in explicitly ecological or other innovative tropes, is taking place. What follows is a contribution to these rethinking efforts in many disciplines.

Overview of the book

A wide-ranging critique of climate change reveals how a focus on climate can be progressive in some ways and in other ways confining. The thematic diversity of the chapters collected in this book is intentionally broad and draws from multiple perspectives and disciplines. Multiple spatial scales come into play, as does a range of actors and institutions. Also under consideration here are methods widely used to assess climate change and anticipated impacts of related processes.

The book is organized into four sections. In the first, authors step back from familiar narratives about climate change to offer insights from alternative perspectives and thereby establish a stance of critique for the rest of the book. In Chapter 2, Leigh Glover argues that the dominant approach to addressing climate change is based on the very same mindset and understanding that created the situation in the first place. In considering climate change as a problem that can be adequately measured and resolved through appropriate market-based solutions, the status quo continues. In Chapter 3, Chris Russill considers how climate change has been represented in the media – the means by which most people are exposed to ideas about this topic – as different modes of warning. But a narrow focus on detection and attempts to link this to hurricane dangers, in particular, has narrowed how climate is discussed, which may in itself be dangerous.

The second section of the book is organized around the theme of justice and conflict. In Chapter 4, Emily Meierding looks at the growing body of scholarly work that ties climate change to human conflict in order to highlight disjunctures between theories and models of conflict developed in other contexts and their (mis)application to the issue of climate change. Paul Routledge considers discourses of climate justice in Chapter 5. Drawing on empirical work in Bangladesh, he assesses trends in climate justice discourses that could be helpful in advancing strategic objectives and practices of social movements and grassroots organizations concerned with climate-related economic and environmental changes. In Chapter 6, Simon Dalby discusses a couple of high-profile attempts to frame climate change as a security threat by the CNA think-tank in Washington and shows that even though high-ranking military figures have been raising the alarm about potential threats to national security, their framing of the issue has been less than successful in international discussions at the UN and, more recently, in domestic US politics.

In the third section, three chapters consider different aspects of science and technology that are related to climate change. In Chapter 7, Shannon O'Lear examines the roots and implications of dominant scientific approaches to interpreting climate change. She considers how geopolitics as usual constrain what kinds of scientific information "count" towards policy-making. In Chapter 8, Thilo Wiertz makes the case that the promotion of climate engineering reflects a management approach designed to naturalize a globally uneven distribution of power. He argues that such a focus on technology actually prevents meaningful change in the status quo. In Chapter 9, James Smith and Shaun Ruysenaar consider how the promotion of biofuels as a response to climate change relies on an insufficient technological understanding of the impacts and emissions of biofuels. They highlight how framing biofuels as "the answer" overlooks numerous important governance and knowledge issues that have quickly turned the promise of a climate "solution" into a troublesome and controversial governance problem.

The final section considers policy and actor responses to climate change at multiple spatial scales. In Chapter 10, Andrew Szasz poses possibilities for new forms of climate activism in the United States. He considers the military, the intelligence community, the insurance industry, and some Christian churches as institutional actors that may have potentially significant impacts on US climate policy and public opinion. To do so will require that they reframe climate change as a matter that relates to their core interests rather than as a peripheral environmental concern. In Chapter 11, Kevin Grove argues that techniques of community-based disaster management and catastrophe insurance shift the balance of power in favour of justice. He considers how disaster resilience can be seen both to reinforce and to resist dominant forms of government control. Angela Oels, in Chapter 12, challenges arguments about climate refugees and the focus on intervening to "save victims" of climate as well as the advocacy of resilience in the face of change. These framings of climate shift the focus from the causes of climate change and as such divert attention from what is most important.

Finally, in Chapter 13, we consider how the critical positions of these authors demonstrate framings of climate change that interrogate dominant narratives and their limited conceptualizations of space and power. Rather than continuing to frame climate change within the current institutions that emerged in the context of state-centred geopolitics and industrial globalization, the collection as a whole considers how we might rethink our structures of governance, practice of politics, spaces of economic activity, science, and scholarship within the context of our altered ecological condition.

References

Agarwal, A. and Narain, S. (1991) *Global Warming in an Unequal World: A Case of Environmental Colonialism*, New Delhi: Centre for Science and Environment.

Agarwal, A. and Narain, S. (1998) "Global warming in an unequal world: A case of environmental colonialism," in K. Conca and G.D. Dabelko (eds) *Green Planet Blues*, Boulder, CO: Westview Press, pp. 157–60.

Barnosky, A.D. et al. (2012) "Approaching a state shift in Earth's biosphere," *Nature*, 486: 52–8.

Braun, B. and Whatmore, S. (eds) (2010) *Political Matter: Technoscience, Democracy and Public Life*, Minneapolis, MN: University of Minnesota Press.

Chalecki, E. (2013) *Environmental Security*, Santa Barbara, CA: Praeger.

Crutzen, P.J. (2002) "Geology of mankind," *Nature*, 415(6867): 23.

Dalby, S. (2009) *Security and Environmental Change*, Cambridge: Polity.

Dalby, S. (2014) "Environmental geopolitics in the twenty-first century," *Alternatives: Global, Local, Political*, 39(1): 3–16.

Goffman, E. (1974) *Frame Analysis: An Essay on the Organization of Experience*, Cambridge, MA: Harvard University Press.

Hommel, D. and Murphy, A.B. (2013) "Rethinking geopolitics in an era of climate change," *GeoJournal*, 78: 507–24.

Hulme, M. (2009) *Why We Disagree about Climate Change*, Cambridge: Cambridge University Press.

Mayer, M. (2012) "Chaotic climate change and security," *International Political Sociology*, 6(2): 165–85.

O'Lear, S. and Gray, A. (2006) "Asking the right questions: Environmental conflict in the case of Azerbaijan," *Area*, 38(4): 390–401.

O'Lear, S., Briggs, C.M., and Denning, G.M. (2013) "Environmental security, military planning and civilian research: The case of water," *Environment*, 55(4): 3–12.

Onuf, N. (2007) "Foreword," in M.E. Pettenger (ed.) *The Social Construction of Climate Change: Power, Knowledge, Norms, Discourses*, Burlington, VT: Ashgate, pp. xi–xv.

Sarewitz, D. (2004) "How science makes environmental controversies worse," *Environmental Science and Policy*, 7: 385–403.

Steffen, W. et al. (2011) "The Anthropocene: From global change to planetary stewardship," *Ambio*, 40: 739–61.

Stiglitz, J.E. and Kaldor, M. (eds) (2013) *The Quest for Security: Protection without Protectionism and the Challenge of Global Governance*, New York: Columbia University Press.

Strauss, S. and Orlove, B. (2003) "Up in the air: The anthropology of weather and climate," in S. Strauss and B. Orlove (eds) *Weather, Climate, Culture*, New York: Berg, pp. 3–14.

Szasz, A. (2007) *Shopping Our Way to Safety*, Minneapolis, MN: University of Minnesota Press.

Terhalle, M. and Depledge, J. (2013) "Great-power politics, order transition, and climate governance: Insights from international relations theory," *Climate Policy*, 13(5): 572–88.

Trenberth, K. (2012) "Framing the way to relate climate extremes to climate change," *Climatic Change*, 115: 283–90.

Vinthagen, S. (2013) "Ten theses on why we need a 'Social Science Panel on Climate Change,'" *ACME: An International E-Journal for Critical Geographies*, 12(1): 155–76.

Wang, L., Gu, M., and Li, H. (2012) "Influence path and effect of climate change on geopolitical pattern," *Journal of Geographical Sciences*, 22(6): 1117–30.

Webersik, C. (2010) *Climate Change and Security: A Gathering Storm of Global Challenges*, Santa Barbara, CA: Praeger.

World Bank (2014) *Turn Down the Heat: Confronting the New Climate Normal*, Washington, DC: The World Bank.

2
POSTMODERN INTERPRETATIONS

Leigh Glover

Introduction

International efforts to address climate change are in a deep malaise. After initial success in reaching a global agreement to cut greenhouse gas (GHG) emissions in the UN Kyoto Protocol in the mid-1990s and the easy and often unintended emissions reductions achieved over the following years, it now appears that the low-hanging fruit has been harvested. Global emissions remain sufficiently high to guarantee dangerous interference with the climate, a successor agreement to the 1996 Kyoto Protocol will not take effect until 2020 (if international agreement can be reached), nations with emission reduction targets account for only a fraction of total emissions, and the extension to the Kyoto Protocol targets saw several nations pull out of the agreement. This slow progress towards international agreement on emissions reductions is a concern if the aim is to keep future global warming below a two-degree-Celsius increase above the pre-industrial level. A greater cause for pessimism, however, is a pervasive outlook that future progress is not assured, while the need for action grows ever more urgent. Fear seems to have the upper hand over hope.

As UN Environmental Programme (UNEP) Director General Achim Steiner stated:

> The challenge we face is neither a technical nor policy one – it is political: the current pace of action is simply insufficient. The technologies to reduce emission levels to a level consistent with the 2°C target are available and we know which policies we can use to deploy them. However, the political will to do so remains weak.
>
> *(UNEP 2013: x)*

If UNEP is correct, and it is far from alone in holding such a view, then it might be thought that there was official interest in diagnosing the problem of politics. Yet, this is far from the view of the United Nations Framework Convention on Climate Change (UNFCCC) bureaucracy, their technical advisers on the Intergovernmental Panel on Climate Change (IPCC), or the majority of commentators. Rather, the common view is that the solution to the problem is to continue the existing model of international negotiations.

This chapter takes up this challenge and considers the problem of climate change and the responses to the issue of reducing GHG emissions as problems of modernity. In essence, it explores the idea of modernity as providing a meta-framing of climate change.

Climate change crisis as a product of modernity

In many respects, once the Industrial Revolution began, the eventual loss of many global environmental values was essentially assured, even though the resources and values that would be at risk could not be predicted in the 1800s. Indeed, even in the contemporary era, the environmental consequences of many current activities cannot be known with certainty – nor, indeed, the future consequences of current efforts to repair or limit lost values. Several basic historical attributes of the social and technological systems under industrialization make global ecological loss a certainty, including:

- industrialization as a process of resource consumption and waste production;
- industrialization as an enabler of increasing human carrying capacity, especially through agriculture, allowing greater population growth;
- ignorance of most larger-scale and longer-term ecological effects of resource harvesting and pollution;
- social and political systems, including scientific and technological advances, that promote the growth and scale of industrialization;
- widespread industrialization of all facets of productive activity and the rise of mass consumption; and
- economic and political systems that reward growth in production and consumption.

(Ponting 1991)

With such features in place and in the absence of any major countervailing influences, the appropriation and consumption of privately owned resources, appropriated common-pool resources, and common-pool resources, it follows that broad-scale consumption can be otherwise constrained only by biophysical limits (York et al. 2003). That the proportion of the earth's productive capacity appropriated for human use would inevitably increase to the point where its systems would be degraded was ensured.

Given the role and scale of exogenous sources of energy in industrialization (Dukes 2003), it is perhaps not surprising that a by-product of fossil fuel use would turn out to be a cause of a global ecological crisis. Perhaps it is worth remembering in our debate over climate change that it is but one, albeit the largest, of many such similar debates over a wide array of global environmental values being lost and ecological risks being created. Had fate taken a few small, different turns, we might well have had a different set of ecological crises to address, such as those arising from nuclear war or genetic engineering.

Many voices expressed concern and opposed the onslaught of industrialization, but these were primarily concerned with local and regional issues. General ignorance of the global ecological implications of industrialization did not end until the arrival of environmentalism in the 1960s. Given the list of scientific achievements since the Industrial Revolution, it is instructive how long it was before the security of the foundations of human ecology was considered and found to be threatened. As it now turns out, given the time lags in the ocean–atmospheric system, the effects of GHG emissions will be influential over the next several millennia (IPCC 2007b).

Implicit in the Industrial Revolution is the "classic" modernity of the nineteenth century. As a historical milieu, modernity is understood from a plethora of perspectives, including political, social, philosophical, cultural, and artistic. In addition to these familiar social science perspectives, there is the more recent effort to understand modernity as a condition of the social relationship with the environment, and that necessarily draws on an understanding of ecology. Ecological losses threatened the viability of contemporary modernity, prompting the reflexive response in sustainable development (and ecological modernization) marked by faith in objective science, a universal ideology, neoliberal governance processes, and an environmental management approach to policy. These aspects of modernism shaped an understanding of global environmental crises, most prominently global climate change.

Responding to climate change with modernity

It is taken as given that sustainable development is the foundation for dealing with global climate change. Through the values, ideology, language, knowledge, procedures, and institutions of sustainable development, the international community under the UN has framed the problem of climate change. Ever since its famous Conference on Environment and Development (UNCED) in 1972, the UN has made sustainable development central to its activities, both in terms of its constituent agencies dealing with environment and resource issues and in its international agreements for environmental protection, and climate change is no exception.

For example, framing issues are addressed in the IPCC's Fourth Assessment Report, which deals with the mitigation of climate change, where it considers "the relationship between sustainable development (SD) and climate change and presents a number of key concepts that can be used to frame studies of these relationships" (IPCC 2007a: 121). These key concepts include decision-making, risk and uncertainty, costs and benefits, vulnerability and adaptation, equity, and technology. By explicitly beginning with sustainable development as an overarching approach, these more procedurally oriented framing issues fit within that larger framing device. Effectively closed off from the panel's view of framing issues are the "upstream" issues relating to the accommodation of economic growth with the inclusion of the values of environmental protection.

Within the sustainable development paradigm there are several different schools of thought, including green consumerism (consumer choice that favours products with limited environmental impact), environmental economics (an accounting approach that considers the costs and benefits of preventing damage to the environment), and eco-technology (environment-friendly technologies or approaches to minimize the ecological impacts of human activity). As the IPCC (2007a) has shown, each of these approaches can point to successes in reducing emissions; and, for their advocates, further and more substantial successes are possible.

Underlying the various interpretations of sustainable development is modernity, with environmental protection being one of many responses throughout the course of modernity to external and internal challenges. As an enormous literature now demonstrates, sustainable development is a "big tent," covering a wide array of views, but very few commentators have seen it as antithetical to modernity. Success for sustainability is seen as success for modernity. Conversely, the failure of sustainable development exposes modernity to the challenges of ecology.

At the national and international scales, when dealing with the crisis of global climate change, we have taken up the outlook, values, and tools of modernity. Modernity's response to climate change has three defining features: global management of the climate system largely through the knowledge systems of science and technology; the application of liberal-democratic governance through suitable institutions at national and international scales; and the management of the natural world for human purposes (Glover 2006). There is dependency between these features; each element does not necessitate the others, but the ends and means of knowledge, governance, and climate management tie together.

First, there is the goal of managing the global climate system. In effect, the UNFCCC and the current Kyoto Protocol are explicit management documents. This management has as its international goal the "stabilization of greenhouse gas concentrations in the atmosphere at a level that would prevent dangerous anthropogenic interference with the climate system" (under Article Two: "Objective of the Framework Convention"). (Under the UN's Copenhagen Accord, arising from its fifteenth Conference of the Parties (COP15) meeting, "dangerous interference" is interpreted as global warming of more than two degrees Celsius above the pre-industrial average.) Such an intention places the global climate system under control. It is not just the intention to control nature that denotes this goal as a feature of modernity; it is also the scale and reach of this control. Human existence has depended on controlling local natural phenomena for human needs: that is, "food, shelter, and clothing." Controlling global ecological systems to meet human needs is an escalation and scale and type of environmental control that could be contemplated only in contemporary times.

Second, there is the use of science and technology for understanding and manipulating the natural and social realms in fulfilment of the goals of climate management. Within the international climate change regime, "science and technology" is synonymous with knowledge. Under the UNFCCC, the IPCC has the role of advising the UN and performing the function of providing knowledge for progress. Its periodic reports are the standard reference documents on the subject. Modernity is identified with the production and use of scientific knowledge, featuring positivism, empiricism, universalism, and the quest for social progress (Toulmin 1990). Control of nature is achieved through technology and science under modernity; it is a technical project.

Third, there is the governance of climate change, which under the UN and within the majority of developed nations has followed the neoliberal variant of liberal-democratic governance through the agency of suitable institutions at the national and international scales. More specifically, this governance seeks the sustainable development goal of conventional economic growth and development and environmental protection through ecological modernization that seeks to internalize the externalities of the costs of carbon emissions, and by creating national and international carbon markets, thereby fostering an emerging "climate/carbon capitalism." Essentially, this commoditization of carbon was endorsed by a wide array of stakeholders, with many environmental groups lending their support to neoliberal champions of free-market environmentalism.

Under the Kyoto Protocol, the UN has promoted the market-based instrument of emissions trading around the world. This has been adopted most notably by the European Union. Establishing global emissions trading has proved immensely difficult, however, due to such factors as the complexity of the system, the absence of a single controlling authority, the problems of coordinating diverse national approaches and practices, and resistance from vested interests, notably energy and related industries. Although the established economic elite took quite a few years before it accepted this market-based instrument, this did eventually happen, as is evidenced by Sir Nicholas Stern's endorsement of emissions trading and his characterization of climate change as a massive market failure (Stern 2009).

Progress in the contemporary response to climate change

According to the IPCC's Fifth Assessment Report, total anthropogenic GHG emissions have continued to increase continually from 1970 to 2010, with greater decadal increases in recent decades (IPCC 2014b). This emissions growth is expected to continue, so that, without mitigation, global (mean surface) temperature by 2100 will be 3.7–4.8°C higher than pre-industrial levels (although the range is 2.5–7.8°C when climate uncertainty is included; IPCC 2014b: 9).

As to the success of the UNFCCC regime, the IPCC (2014a: 6) concluded that "the overall level of mitigation achieved to date by cooperation appears inadequate to achieve this goal [of limiting global warming to two degrees Celsius above pre-industrial levels]." It found that the Kyoto Protocol's effectiveness "has been less than it could have been," which it explained as resulting from "incomplete participation and compliance" by the developed nations and claims for emissions reductions that "would have occurred without the Protocol in economies in transition." It added, as all followers of the issue know, that the non-Annex I nations (i.e. developing nations) have rapidly growing emissions and are unregulated by the Protocol.

Extending the Protocol under the Doha Amendment from 2013 to 2020 (the second commitment period) revised national emissions reduction targets. However, the effectiveness of the post-Kyoto Protocol agreement process has been below official expectations as successive international meetings have struggled to reach agreement. Despite consensus around the dangers of a high degree of future warming (see, for example, World Bank 2012), the "emissions gap" between current commitments and the level of reductions needed to prevent a warming of two degrees Celsius remains a problem (UNEP 2013). Full implementation of the second commitment period obligations would cut global emissions by five gigatonnes of carbon dioxide equivalent ($GtCO_2$-e), but that is still between eight and twelve $GtCO_2$-e short of what is required to close the emissions gap (UNEP 2013).

So far, carbon pricing rules have been implemented by only about 40 nations and about 20 sub-national jurisdictions that cover about 12 per cent of global annual GHG emissions (World Bank 2014). Full emissions trading involves less than even these totals suggest. For an international agreement and a policy instrument that were intended to have international coverage, progress is far below the original goals.

On emissions trading – the centrepiece of the UN response to mitigating GHG emissions – the IPCC is cautious, if not equivocal. It has found that market mechanisms "could reduce overall emission costs," while pricing carbon "could create incentives ... [and] could include economic instruments, government funding and regulation" (IPCC 2007a: 19). Rather than finding market-based environmentalism that allows some flexibility and negotiation superior to more rigid regulatory instruments, it suggested that there is a "wide variety of national policies and instruments available to governments," that their applicability depends on national circumstances, and that "experiences and sectors show there are advantages and disadvantages for any given instrument" (IPCC 2007a: 19).

Almost on the final page of its Fourth Assessment Report on GHG mitigation, the IPCC (2007a: 795) poses six questions:

1   Why has the application of policies been so modest?
2   Why is the global community not on a faster implementation track?

3   Why have – at the very least – hedging strategies not emerged in many more countries?
4   Is the scale of the problem too large for current institutions?
5   Is there a lack of information on potential impacts or on low-cost options?
6   Has policy-making been influenced by the special interests of a few?

These are certainly pertinent questions, to such an extent that humanity's future on the planet would seem to depend on finding the correct answers to them. Indeed, one might have thought that the IPCC had a responsibility to be addressing, rather than merely posing, them. The brief answers it does provide are telling, however. For instance, it suggests that GHG emissions changes can reflect non-climate change policies and governments' desire for cheap energy, economic growth, the competitive global economy, and the influence of special interests. This is about as close to political economy that the IPCC has ever got in explicit form.

Modernity's failures in responding to climate change

These "last-page" IPCC reflections are emblematic of the contemporary impasse on climate change, wherein we recognize that there have been insufficient successes due to some major social and economic roadblocks and simultaneously endorse continued incrementalism without an overall understanding that these itemized problems might constitute elements of systemic failure.

Although there are many theories and ideas in circulation relating to how to repair the failings of the national and international climate change responses, here we take the position that the fault lies at the foundation of these responses – namely, the problem of modernity. Decisions to redouble modernization efforts to address GHG emissions are justified on the basis that the efforts to date have not been as effective as they must be. These decisions are, in effect, efforts to repair the repairs of ecological modernization. However, they are founded on the belief that further ecological modernization can succeed. Yet, it may be that the cure for modernity is not modernity and that such efforts are contradictory. Three arguments are presented on these environmental contradictions of modernity concerning climate change based on the three defining features outlined above.

Contradictions of global climate management

There have been few greater expressions of modernity's ambition and hubris than aspiring to control the global climate system, for, having conducted the "vast and uncontrolled experiment" with global climate only to learn in the wake of the experiment what risks have been taken and still be uncertain as to what has occurred, the next response is to assume that it will be amenable to manipulation through agreements established by a conflicted set of nation states and global corporations. Simply put, the ambition of global climate control is flawed in both (political) means and (ecological) ends.

That climate change is a global problem requiring a global response is the mantra of the UNFCCC, a proposition enshrined in the Kyoto Protocol and given the form of differentiated national responsibilities for taking action to limit GHG emissions. Politically, the device of "common but differentiated responsibilities" was sufficient to gain international agreement for the Protocol by having developed nations accept GHG emissions targets (these nations were listed under Annex I in the Protocol). But, in practice, the Annex I/non-Annex I split of nations secured only modest (at best) reduction commitments from the former. Any rigorous association between accumulated and current national emissions was reduced further when individual developed nations negotiated their own targets under the Protocol, including a small group that negotiated *increases* in their emissions. However, inertia in these ongoing negotiations has led to a situation in which the original developed nation emission targets form the foundation of current targets. As a construct of a "global response," this key aspect of the Protocol has proved to be particularly unconvincing in practice.

Irrespective of whether the principle of common but differentiated responsibilities is ethical, it has played a crucial role in the political and practical failings of the framework convention. It has, for example, meant that China has not been obliged to meet any emissions reduction target – a fact that gave rhetorical cover to the United States' decision not to sign the Protocol. China is the world's largest emitter (with India now fourth), and future emissions growth will likely be concentrated in developing nations. We now enter the third decade of the UNFCCC without an effective global agreement. As things stand at the time of writing (late 2014), the Protocol targets for developed nations have been extended until a new agreement is reached, planned for 2015, and the new targets to come into effect in 2020. Whether this new agreement will extend the common but differentiated principle is unclear, but recent Conferences of the Parties suggest that developing nations will lobby for its continuance.

Symbolically, the differentiated responsibilities approach proved to be a weak platform for negotiations following the 1996 Protocol. As a strategy, the approach was a thin recognition of historical emissions; tactically, it was of questionable value, since it arguably made continuing negotiations less productive. Put crudely, this vast enterprise in high-handed diplomacy and base national politics has brought us an emissions reduction of a few per cent from the Annex I group (and even that handily included the emissions savings that resulted from the economic collapse of the Eastern Bloc).

Although the global politics research community has published widely on the problems of the UNFCCC, the conventional nexus between the role of all nation states and the formulation of climate change as a global problem is common. This role for nation states is present in studies identifying the emergence of shared norms between nations over environmental threats in the post-realist school of international relations, included in which would be recent interest in game theory explanations of nation state strategies in negotiations. In contrast to the plethora of diagnoses, international relations does not seem to have come up with many plausible solutions to the impasse in negotiations.

Hoffman (2011) and Harris (2013) are among the few to question the viability of continuing with the nation state as the unit of cooperation. As Harris (2013: 63) concludes,

> the fundamental pattern of international climate politics has changed little since the 1980s ... The failure of international negotiations to achieve agreements that will do more to avert catastrophic climate change – their stated objective from the outset – can be largely attributed to the cancer of Westphalia. The norms that serve as the basis of international relations have proved to be more powerful than the nascent and evolving norm of environmental stewardship.

For Harris, this "cancer of Westphalia" is the nation states' narrow self-interest and short-term perspective, which prevent them from meaningfully addressing climate change through international agreements. It appears that despite the vast apparatus of the UNFCCC, and the length of time taken to develop an effective successor to the Kyoto Protocol, such an agreement will not and cannot be reached.

As to whether the global climate can and should be managed, there are grounds for questioning this most basic objective of the climate change response. As Timothy Luke states (2008: 128–9): "To even speak of 'greenhouse gases' already implies the earth can now be best understood as an essentially built environment, a human–machine hybrid, or a vast artifice ironically fabricated by wastes, by-products, or effluents." Many times in the fracas over legitimate climate change science and the industry-sponsored climate sceptics, the latter group and their tactics have been likened to those employed by "big tobacco" to discredit the science linking smoking and cancer. A similar metaphor can be applied here if we consider the differences between the public health responses to the dangers of tobacco smoking and drinking alcohol. A key difference between these human health risks is that there is no safe level for smoking, whereas light intakes of alcohol are generally assumed to have no serious health implications. In the case of limiting GHG emissions to prevent dangerous interference with the climate system, we have treated them as akin to alcohol consumption by presupposing that we know the safe levels. Here, the problem is that our two-degree-Celsius warming goal is really nothing more than a scientifically informed guess. It may be that GHG emissions and climate change are actually more akin to smoking, and that there is no "safe" level of emissions.

Turning the argument around, could it be argued that dangerous interference with the climate system has already occurred? Is limiting warming to two degrees Celsius since the start of the Industrial Revolution "safe"? According to Blunden and Arndt (2014), the Mauna Loa observatory recorded an atmospheric carbon dioxide level of 400 ppm for the first time in 2013, thereby breaking through a symbolic barrier (of sorts). For context, the level did not exceed 300 ppm at any point during the 800,000 years prior to the twentieth century. Given what has happened recently in an array of climate vectors – surface temperatures, sea surface temperatures, sea levels, northern hemisphere snow cover, Arctic sea-ice volumes, and mountain glaciers – it is not difficult to construct an argument that dangerous climatic interference has already occurred.

There is something odd about the logic of arguing for set emission targets and limits to control the extent of damage to a system whose behaviour we cannot confidently predict – and the effects will last for millennia (see, for example, Solomon et al. 2009). A particularly worrying aspect of the climate system's behaviour is the possibility of abrupt, major climate change when various thresholds are passed (see, for example, Lenton et al. 2008). Most scenarios of future climate change, such as those of the IPCC, are based on slow, gradual shifts. Abrupt climate change would be more difficult for adaptation and more socially and environmentally disruptive than gradual change, but it is a prospect that remains elusive in the modelling of future climate. Not only is the notion that climate may be managed a dubious one, but it is compounded by the use of temperature as the political symbol and metric of global climate change. Of course, temperature is but one measure of climate.

Contradictions in scientific and technological solutions

As the UNFCCC follows the dictates of sustainable development, it gives pride of place to the transformative role of technology, and especially "eco-efficiency." Technology is sought to provide solutions created out of social relations, such as the selection and use of existing technology, whereby existing social relations and relationships with the natural world can be left largely undisturbed. In effect, this approach is "post-political," to use Swyngedouw's (2010) term, in that the social and economic factors that contribute to its direction, choices, and priorities of knowledge creation and distribution are not subject to self-examination. Prioritizing technology, that is, forecloses discussion of other approaches to human–environment relations. For example, van der Sluijs et al. (2010) argue that the IPCC's model of reaching (scientific) consensus politicized its findings (because scientific and political dissent are essentially neglected), thereby inhibiting a broader debate to the detriment of democracy. In any event, a number of aspects of the use of scientific and technological knowledge are worthy of closer scrutiny.

Replacing the fossil fuel energy system with renewable energy dominates this technological challenge. Science and technology provide the answer to the energy foundation of the Industrial Revolution – namely, the search for low-cost and abundant exogenous energy for industrial activity. By approaching this challenge as a technological problem, the underlying social and economic systems embedded in mass consumption and the "treadmill of production" – the essential logic of the fossil fuel system – are undisturbed. Abundant and profligate use of energy fuels industrialization with all its attendant environmental costs, so seeking to substitute another energy source becomes, in effect, only an increase in the efficiency of production, thereby exhibiting high-energy-use "lock-in." High-energy-use economies powered by renewable energy will continue to consume high levels of natural resources and environmental services, thereby continuing the global environmental crisis. Globalized industrialization will continue to degrade the global ecosystems even when powered by renewable energy, albeit possibly at a slower rate.

An ecologically sustainable energy system can be achieved only by minimizing energy consumption. Seeking an energy system transformation through nothing but technological change while leaving the causal social relations intact is self-defeating. Such a way forward to a sustainable energy system is blocked by the contradiction between promoting high-energy-use economies and the need for dematerialization and reduced energy and resource throughput.

One of the major limits of technological solutions to the energy problem of climate change is an inability to address economic growth. As Herman Daly (1996) stated 20 years ago, sustainable growth in a finite world is an "impossibility theorem." Even if greater efficiency in energy consumption finds greater favour, growing global energy use will merely delay the environmental consequences of that growth; today's gains will be eroded by tomorrow's growth. Despite the insights of ecological economists such as Daly, proponents of sustainable development have still failed to acknowledge the flaws in its basic logic. By harnessing technology as a solution to the energy problem, the international community continues its commitment to globalized economic growth and its attendant implications for continued alteration of the planet's ecosystems and atmosphere.

Other barriers undermine the viability of technological approaches to the "energy problem." It is very unlikely that renewable energy technologies will be able to substitute for fossil fuels completely because of the scale of energy required, the costs of some renewable energy sources, the practicality of transporting energy, and the types of energy services that fossil fuels are particularly well suited to provide. Added to this is the length of time that transformation of the energy system seems to be taking, which is at odds with the urgent need to reduce GHG emissions. Because the energy problem has been conceived as a technological (and economic) issue, the essential task of reducing energy consumption and resource throughput in industrial societies is viewed as far less of a priority than finding ways to continue the model of industrialization based on energy systems with cheap, abundant, and portable forms of energy.

Finally, attempted scientific and technological solutions to climate change may worsen the interlocking global environmental crises of human causation. Although it is hard to believe that global geoengineering might ever be considered a potential solution to global warming, as Wiertz shows (Chapter 8, this volume), research in this area has been considerable and remains ongoing.

As might be expected, such projects are often cast as "there is no alternative" imperatives by their promoters. However, having unwittingly and catastrophically experimented with the global climate system to the apparent detriment of global ecology, there would seem to be no reasonable ethical or practical grounds for *deliberately* conducting speculative experiments on the global climate and its associated systems (Hulme 2014).

Contradictions of liberal-democratic governance

Clearly, carbon capitalism is possible, as is demonstrated by the existing carbon trading schemes, but the ultimate success of this model can be shown only by its ability to transform the global economy. Emissions trading is applied as the solution to the problem of climate change as a "market failure." Such an instrument of climate management has governance implications. As Redclift (2012: 26; emphasis in original) concludes:

> The characterisation of climate change as a "market failure" immediately offered economists, businesses and government a lifeline. Rather than necessitating expensive and comprehensive restructuring in new systems of provision, or even reduced volumes of production and consumption, Stern's neoclassical view was that sustainability could be delivered through *increased* consumption of particular kinds of products simultaneously. Feeding the economy has come to typify the mainstream environment and consumption discourse.

For environmental critics of capitalism, global carbon markets are inherently contradictory. O'Connor's (1988: 25) account of the "second contradiction of capital" holds that "An ecological Marxist account of capitalism as a crisis-ridden system focuses on the way that the combined power of capitalist production relations and productive forces self-destruct by impairing or destroying rather than reproducing their own conditions." From this it follows that governance that seeks to reconcile capitalism with environmental protection is flawed, essentially because capitalism cannot be constrained in this way. Kovel (2007) identifies three specific causes:

1   Capitalism tends to degrade the conditions of its own production.
2   Capital expands without end in order to flourish.
3   Capitalism leads to increasing inequity so that the world system is chaotic and is incapable of addressing the ecological crisis.

As the studies of successful common-pool resource management systems by Elinor Ostrom (1990) and others have shown, the stocks and flows of natural resources cannot be protected in free markets. They require a specific set of rules and social institutions that serve to constrain the market incentives.

There are other contradictions within the liberal-democratic governance of climate change, too. Basic political economy suggests, for instance, that transforming global energy systems from fossil fuels to renewable energy is not facilitated by political and economic institutions strongly influenced by vested and entrenched interests in the conventional energy system. One does not have to look very far in the accounts of national climate change politics to find evidence of success for the opponents of taking effective action on climate change. Although sustainable development has shown how state and corporate interests can be brought together to work in concert for environmental protection, it also shows that there are limits to the reform that is possible under status quo arrangements. Nations with strong economic interests in the energy sector nearly always retain various public subsidies to key industries and firms while simultaneously promoting action on climate change, often with comparatively much lower levels of investment. These subsidies tend to be so large that even the International Monetary Fund has called for their end; in 2013 it estimated that the global fossil fuel subsidy stood at US$1.9 trillion and that its cessation would reduce global GHG emissions by between 1 and 2 per cent (IMF 2013).

Turning to the instrument of carbon trading itself, for those rejecting the preceding arguments as facile or misguided and insisting that "there is no alternative" to emissions trading, the performance and prospects of this must be essentially faultless, given what is at stake. However, any such confidence appears misplaced as, on both empirical and conceptual grounds, emissions trading seems deeply flawed and systemic problems make any future success unlikely (Lohmann 2008).

Without doubt, the sheer complexity of what global carbon capitalism proposes is daunting; it is already far beyond the understanding of individuals and experts in the field who have specialized knowledge in specific aspects of the system. Such complexity has stifled progress to date and, given that the system will only become more complicated with further development, this must surely count against any prospect of future success. Anyone doubting this complexity might acquaint themselves with some of the relevant documentation from the UNFCCC processes. Complexity is not necessarily a fault in itself, but it generates serious impediments to implementation and operation.

For democratic nation states, "getting the message out" has been difficult, with ongoing confusion about how emissions trading works. Furthermore, unfamiliarity with this novel instrument has made it easy prey to vested interests that have stirred wider confusion and concern. Contrary to the ideal that carbon capitalism will broadly engage all consumers, for the most part emissions trading systems are and will continue to be the territory of experts and specialists, creating a new managerial class within the financial industries and state bureaucracies. And, of course, the experts can get it wrong, as was revealed in the price crash of 2013 during the first phase of the European Union Emissions Trading Scheme (EU ETS), the world's largest international system for trading allowances for carbon emissions.

Second, as with so much else of the UNFCCC, there is a technical imprimatur over questions of values and politics. Far from the design and operation of emissions trading following a neutral template, every major design aspect requires subjective decisions and evokes political values. Allocating emission permits (such as provided gratis or auctioned), the coverage of direct and indirect emissions, the type of target (such as relative versus absolute), industry sector coverage and exemptions, permit banking and borrowing, monitoring, and enforcement all entail decisions with distributional and equity implications. For example, there were generous free permit allocations in phase I of the EU ETS, to the extent that some nations allocated more permits than actual emissions as a consequence of (unreliable) forecasts. Furthermore, these decisions are unlikely to be subject to the sorts of checks and balances that might be applied to other major public policy initiatives in democratic nation states. At the international and global level, these institutions will be removed from national political accountability.

Third, there is the matter of legitimacy. One of the more striking aspects of the carbon markets is that, notwithstanding all of the research and advocacy, a key component – namely, ensuring the legitimacy of purchased carbon offsets – has been poorly and haphazardly undertaken. A number of studies have exposed this weakness. For example, Rogers (2010: 172) opined: "The offset mechanism rests on two rather wobbly legs – 'baseline' and 'additionality.'" This uncertainty arises from the substantial supposition that goes into establishing the baseline level of emissions in many projects, a problem that is exacerbated in the almost impossible task of assuring additionality (i.e. whether a specific project would have occurred without offset funding). While corporate carbon offsets under the Kyoto Protocol are subject to an array of controls, the voluntary offset sector – as might be used by you or me – is at the raw end of capitalism. Rogers (2010: 153) found "Offset brokers are not compelled to meet any standards, have no required inspections, project approvals, or reviews, and no obligatory follow-up assessments to ensure the efficacy of the carbon mediation." Environmental and social losses in developing nations arising from offset projects have been documented (see, for example, IPCC 2007a). Other aspects of legitimacy are "carbon leakage," wherein emission sources (and products) are transferred out of the jurisdiction of emissions trading, typically to another country. Examples of windfall profits resulting from the free allocation of permits to existing firms do little to foster confidence in the legitimacy of emissions trading in practice.

Fourth, there is the disturbing possibility that an established trading scheme will be both profitable and ineffective in lowering emissions. Conflicts between ecological protection and profit maximization, with the victory of the latter, could occur in a number of ways. By following price signals of the carbon market, at certain times and under certain conditions, some firms will decide that continued GHG pollution is the cheaper option. While carbon pricing *can* provide incentives for incremental improvements to reduce GHG emissions, it may also serve to perpetuate existing practices by delaying more fundamental changes. Where the major firms in industries can pass the costs of emission permits on to consumers, the existing system of producing emissions is normalized, leaving individual consumers with the problem of finding greater efficiencies and alternative services. As financial commodities, carbon permits could be combined with other financial commodities for trading and speculation and could be liable to global crises and potential price collapses. Fundamentally, using market-based policy instruments bolsters the global market system and the rationale of conventional economics – namely, that the environment should be protected only to the extent that the existing economic system is not disrupted. This illustrates the basic contradiction between ecology and conventional economics.

Postmodern environmentalism

Modernity can be understood as an environmental phenomenon, as distinct from a cultural or social condition of development. Modernity's relationship to ecology, as demonstrated in the climate change crisis, is marked by contradictions. Globalized models of industrialization have come to undermine the ecological systems on which the model is based. In turning to the tools of modernity for a way out of the crisis, the limits of the ecological conditions of modernity are exposed. Meaningful progress in addressing the crisis by controlling GHG emissions seems to have stalled. As such, we have reached the condition of postmodern environmentalism; social action is needed to limit the damage caused to global ecosystems, including the global climate system, but this action cannot be based on industrial modernity.

Within the response to climate change under modernity, we have identified three nested hierarchies: the social domination of nature through aspirations to global environmental management for human ends; the domination of science and technology over social relations; and the domination of vested economic and political interests within society. In considering reframing the response to climate change, the state-centred system is unlikely to be the most appropriate structure through which to question these established hierarchies. Social relations will be the necessary starting point to shape an ecological geopolitics from which to question these systems of domination.

Although GHG distribution in the atmosphere may be relatively even or homogeneous, the contributing factors are not. Climate change is a reflection of inequalities of wealth and well-being, of consumption levels, of resource use, and of waste generation, which are generated through uneven policy-making, corporate activity, methods of accounting, valuation and financialization, and marketing, not to mention the planet's inherently uneven distribution of resources and ecosystem services. Our collective success in addressing climate change may well depend on the extent to which these forms of oppression can be rejected and overturned.

References

Blunden, J. and Arndt, D.S. (eds) (2014) "State of the climate in 2013," *Bulletin of the American Meteorological Society*, 95(7): S1–S257.

Daly, H.E. (1996) "Sustainable growth: An impossibility theorem," in H.E. Daly and K.N. Townsend (eds) *Valuing the Earth: Economics, Ecology, Ethics*, Cambridge, MA: MIT Press, pp. 267–73.
Dukes, J.S. (2003) "Burning buried sunshine: Human consumption of ancient solar energy," *Climatic Change*, 61(1–2): 31–44.
Glover, L. (2006) *Postmodern Climate Change*, New York and London: Routledge.
Harris, P.G. (2013) *What's Wrong with Climate Change and How to Fix It*, Cambridge: Polity.
Hoffman, M. (2011) *Climate Governance at the Crossroads: Experimenting with a Global Response after Kyoto*, Oxford: Oxford University Press.
Hulme, M. (2014) *Can Science Fix Climate Change? A Case against Climate Engineering*, Cambridge: Polity.
International Monetary Fund (IMF) (2013) *Energy Subsidy Reform: Lessons and Implications*, Washington, DC: IMF.
Intergovernmental Panel on Climate Change (IPCC) (2007a) *Climate Change 2007: Mitigation of Climate Change*, Cambridge: Cambridge University Press.
Intergovernmental Panel on Climate Change (IPCC) (2007b) *Climate Change 2007: The Physical Basis of Climate Change*, Cambridge: Cambridge University Press.
Intergovernmental Panel on Climate Change (IPCC) (2014a) "International Cooperation: Agreements and Instruments," in O.R. Edenhofer, R. Pichs-Madruga, Y. Sonoka, E. Farahani, S. Kadner, K. Seyboth, A. Adler, I. Baum, S. Brunner, P. Eickemeier, B. Kriemann, J. Savolainen, S. Schlömer, C. von Stechow, T. Zwickel, and J.C. Minx (eds) *Climate Change 2014: Mitigation of Climate Change: Contribution of Working Group III to the Fifth Assessment Report of the Intergovernmental Panel on Climate Change*, Cambridge and New York: Cambridge University Press, pp. 1001–82.
Intergovernmental Panel on Climate Change (IPCC) (2014b) "Summary for Policymakers," in O.R. Edenhofer, R. Pichs-Madruga, Y. Sonoka, E. Farahani, S. Kadner, K. Seyboth, A. Adler, I. Baum, S. Brunner, P. Eickemeier, B. Kriemann, J. Savolainen, S. Schlömer, C. von Stechow, T. Zwickel, and J.C. Minx (eds) *Climate Change 2014: Mitigation of Climate Change: Contribution of Working Group III to the Fifth Assessment Report of the Intergovernmental Panel on Climate Change*, Cambridge and New York: Cambridge University Press, pp. 1–33.
Kovel, J. (2007) *The Enemy of Nature: The End of Capitalism or the End of the World?*, London and New York: Zed Books.
Lenton, T.M., Held, H., Kriegler, E., Hall, J.W., Lucht, W., Rahmstorf, S., and Schellnhuber, H.J. (2008) "Tipping points in the earth's climate system," *Proceedings of the National Academy of Sciences*, 105(6): 1786–93.
Lohmann, L. (2008) "Carbon trading, climate justice and the production of ignorance: Ten examples," *Development*, 51(3): 359–65.
Luke, T.W. (2008) "Climatologies as social critique: The social construction/creation of global warming, global dimming, and global cooling," in S. Vanderheiden (ed.) *Political Theory and Global Climate Change*, Cambridge, MA: MIT Press, pp. 121–52.
O'Connor, J. (1988) "Capitalism, nature, socialism: A theoretical introduction," *Capitalism Nature Socialism*, 1(1): 11–38.
Ostrom, E. (1990) *Governing the Commons: The Evolution of Institutions for Collective Action*, Cambridge: Cambridge University Press.
Ponting, C. (1991) *A Green History of the World: The Environment and the Collapse of Great Civilizations*, Harmondsworth: Penguin.
Redclift, M.R. (2012) "Living with a new crisis: Climate change and transitions out of carbon dependency," in M. Pelling, D. Manuel-Navarrete, and M.R. Redclift (eds)

Climate Change and the Crisis of Capitalism: A Chance to Reclaim Self, Society and Nature, London: Routledge, pp. 21–36.

Rogers, H. (2010) *Green Gone Wrong: How Our Economy is Undermining the Environmental Revolution*, New York: Scribner.

Solomon, S., Plattner, G.-K., Knutti, R., and Friedlingstein, P. (2009) "Irreversible climate change due to carbon dioxide emissions," *Proceedings of the National Academy of Sciences*, 106(6): 1704–09.

Stern, N. (2009) *A Blueprint for a Safer Planet: How to Manage Climate Change and Create a New Era of Progress and Prosperity*, London: The Bodley Head.

Swyngedouw, E. (2010) "Apocalypse forever? Post-political populism and the spectre of climate change," *Theory, Culture & Society*, 27(2–3): 213–32.

Toulmin, S. (1990) *Cosmopolis: The Hidden Agenda of Modernity*, Chicago, IL: University of Chicago Press.

United Nations Environment Programme (UNEP) (2013) *The Emissions Gap Report 2013*, Nairobi: United Nations Environment Programme.

van der Sluijs, J.P., van Est, R., and Riphagen, M. (2010) "Beyond consensus: Reflections from a democratic perspective on the interaction between climate politics and science," *Current Opinion in Environmental Sustainability*, 2(5–6): 409–15.

World Bank (2012) *Turn Down the Heat: Why a 4°C Warmer World Must Be Avoided*, Washington, DC: The World Bank.

World Bank (2014) *State Trends of Carbon Pricing 2014*, Washington, DC: The World Bank.

York, R., Rosa, E.A., and Dietz, T. (2003) "Footprints on the earth: The environmental consequences of modernity," *American Sociological Review*, 68(2): 279–300.

3
THE CLIMATE OF COMMUNICATION
From detection to danger

Chris Russill

> What has changed is that our common sense has begun searching for a language to speak about the shadow our future throws.
>
> *(Illich 1989)*

Introduction

Is climate change dangerous?

This question is of relatively recent origin. In 2006, it was forced to popular attention by *An Inconvenient Truth* and the strong association made in the film between climate change and natural disaster. The danger implied by this relationship was dramatized by reference to the catastrophic effects of Hurricane Katrina. Images of the storm's massive eye wall, photographs of destruction, and heart-breaking audio of a person near drowning offered audiences quite visceral materials for imagining danger. The film emphasized this point by depicting a hurricane cloud formed from the smokestack emissions of a coal plant on its promotional material (see Figure 3.1).

The film, of course, is built around Al Gore delivering a climate science lesson, and it dramatizes the difficulties he encounters in conveying his message. There is no question that Gore has a strong understanding of anthropogenic global warming. His account of its detection is fluent and accessible, and credits individual scientists by name. Yet, on the association of climate change and hurricanes, Gore is much more vague, simply hinting that warnings went unheeded, while concluding: "One question that we, as a people, need to decide is how we react when we hear warnings from the leading scientists in the world" (Gore, quoted in Guggenheim 2006).

Gore's question is the right one.

Oddly, we do not hear warnings from scientists in the film. As mentioned above, Gore does name several of them – including Roger Revelle, David Keeling,

32 Chris Russill

FIGURE 3.1 Poster for *An Inconvenient Truth* (2006).

Source: *Inconvenient Truth, An,* Year: 2006, Dir: Guggenheim, Davis, Ref: INC027AL, Credit: Lawrence Bender Prods/The Kobal Collection

and Lonnie Thompson – but they do not appear in the film or describe climate change as dangerous. Gore met numerous other scientists when operating the dominant interface for climate science and public life in the 1980s: the governmental hearing. Throughout that decade, he and his colleagues invited scientists to inform policy-makers. The records of these meetings are remarkably rich and repay careful consideration. Yet, unlike public discourse in the 1960s and 1970s – the heyday of ominous warnings from oceanographers (Maurice Ewing, Wally Broecker) and atmospheric scientists (Reid Bryson, Stephen Schneider) – one finds few if any urgent claims of danger from climate scientists in this culturally authoritative context.

In the 1990s, the relationship between climate science and public life was reconfigured to recognize the priority of the UN Intergovernmental Panel on Climate Change (IPCC) and, to a lesser extent, the US Global Change Research Program (GCRP). These were formal institutions for developing interdisciplinary, scientific syntheses and intended to make climate change science actionable for policy-makers. On this model, science was not reported through panels of scientists offering an array of opinions but generated through a collective and consensus-based voice. The IPCC, in particular, is designed to inform the international policy community convened through the UN Framework Convention on Climate Change (UNFCCC) and to facilitate avoidance of "dangerous anthropogenic interference" with the climate system. This institutionalization of "the" voice of science brought scientific interpretation of climate change within an arena of political management of climate change. Key to this arrangement was a framing of climate change that sits comfortably with the status quo (see O'Lear, Chapter 7, this volume, for further discussion).

The IPCC incorporated the goal of avoiding danger and this goal has shaped how the panel portrays anthropogenic climate change: as certain, undesirable, and potentially risky in its impacts, but *not* dangerous. For the most part, danger was conceptualized in this approach as an effect of changes in global mean surface temperature (GMT), a long-accepted measure for determining global warming; danger would be registered, known, and understood primarily at the global scale through an indicator that permitted only incremental change (given that it is a global average of temperature). See, for example, Figure 3.2, where reasons for concern over climate change are pegged to increases in GMT.

The warning

In 2005, the conventional framing of climate change was upset. James Hansen, a climatologist working at the National Aeronautics and Space Administration (NASA), became frustrated with state-authorized processes and spoke out publicly to warn unambiguously of impending danger. Hansen described global warming as a time bomb and followed with a stark warning of climate danger that criticized the IPCC for improperly communicating the threats faced by humanity:

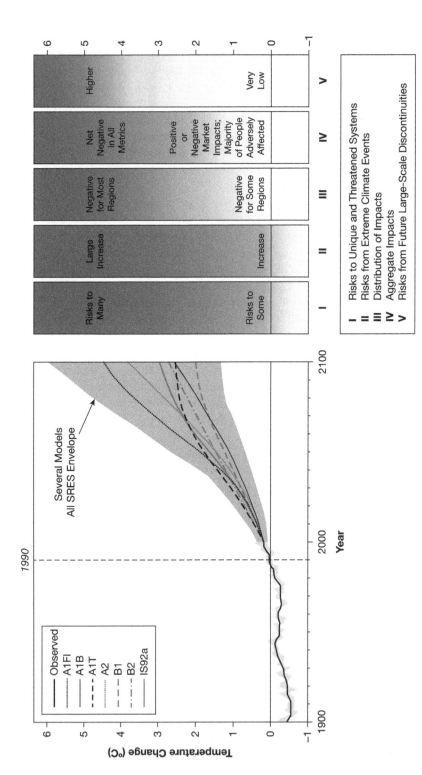

FIGURE 3.2 Reasons for concern over climate change pegged to increases in global mean surface temperature

Source: Figure SPM-2 in McCarthy et al. (2001)

> A critical issue is then: what level of global warming would constitute "dangerous anthropogenic interference"? IPCC uses a burning embers diagram to quantify reasons for concern about climate change. The impression created is that 2 or 3 degrees Celsius warming, relative to the present, is probably dangerous. The burning embers are usually interpreted with a probabilistic approach, which has certain merits. However, I suggest that the burning embers are a fuzzy concept that discourages action, action that is needed urgently, because we are on the precipice of climate system tipping points beyond which there is no redemption.
>
> *(Hansen 2005: 7–8)*

Danger, from Hansen's perspective, had been improperly conceptualized. Yet, he was not merely frustrated with the IPCC or a prominent graph. His warning had the destruction of a major US city as its immediate context. In fact, several investigations into the Hurricane Katrina catastrophe demonstrated that the federal government had failed to respond to clear and persistent warnings of impending danger, and one Federal Emergency Management Agency (FEMA) official, Michael Brown, had attributed the disaster to tipping points that the federal government could not manage (Russill and Lavin 2012).

Hansen, in this context, stated unequivocally that humanity had reached climate change tipping points that promised dangerous consequences if they were crossed. Roughly speaking, tipping points signify non-linear transitions from one stable state to another stable state, transitions that are usually abrupt, irreversible, and challenging to economic, urban, social, agricultural, or other systems. That is why tipping points may be seen as dangerous. When NASA officials told Hansen to refrain from using the word "dangerous," Hansen refused, and the ensuing controversy pushed his warnings abruptly to the forefront of public discourse. The question of determining danger – and of how to respond to scientists' warnings of imminent danger – had jumped from the rarefied pages of specialist journals, Hollywood films, and popular dissent to mainstream discourse and public debate.

Gore, as we know, was making a film about climate change at this time. It seems that he had already arranged for Hansen to assess the scientific accuracy of the slide show featured in *An Inconvenient Truth* before the controversy broke out (Hansen 2006b). At the time of their meeting, Hansen believed climate change was fuelling stronger hurricanes, an idea he had excised from his 2005 "tipping point" warning due to constraints on the time he was permitted to speak (Hansen 2006a). Hansen also believed government officials at the National Oceanic and Atmospheric Administration (NOAA), like those at NASA, were censoring scientific warnings of danger, and he stepped forward publicly to support a strong association between climate change and hurricane intensity.

An Inconvenient Truth amplified these claims considerably, and Hurricane Katrina became an icon and analogue for assessing future danger, if not direct evidence of a spectacular climate change disaster, depending on how one read the film.

The warnings of danger soon proliferated.

Climate scientist Stephen Schneider (2006) repeatedly connected climate change to extreme weather, and particularly hurricane intensity, after Katrina. In 2008 earth systems scientist John Schellnhuber organized an "exhibition of the worst nightmares of climate scientists" to give the fears associated with tipping points a collective outlet (see Schneider and Nocke 2014: 14). Risk society theorist Ulrich Beck (2010) wondered aloud why there was no Red October event. Schellnhuber and Beck called for revolutions.

The surge of warnings offered a more dramatic and alarming depiction of climate change than state governments or the IPCC had offered. A growing number of scientists, including Hansen, had lost faith that the institutional system organized by the UN would prove sufficient for managing – and distributing – the risks and dangers of climate change. While some likely hoped that more urgent warnings would propel the political process forward – an assumption prompting many climate communication experts to recommend the downplaying of threatening or fearful language (see O'Neill and Nicholson-Cole 2009) – others were reframing climate change altogether, often with an emphasis on the challenges that tipping points posed to prevailing governance strategies (Lenton and Schellnhuber 2007). For some, this meant humans were forcing the planet into a different epoch and acting collectively as geological agents (Chakrabarty 2009). A more diverse set of perspectives on climate change and danger had been articulated. The tidy containment of danger as originally envisioned by the creators of the IPCC was no longer in effect.

Why did evidence of the detection of anthropogenic climate change in the late twentieth century not create the political conditions for avoiding dangerous climate change? Important work has been done on this question and diverse hypotheses abound. Scholars have argued that the problem lies in the scientific illiteracy of the public, the wiring of human brains to near-term threats, the bad faith of climate change sceptics, the intransigence of countries dependent on fossil fuel exports, the inability of scientists to speak clearly, the hubris and politics of liberal democracies, the evils of capitalism, the failings of consumer culture, and even the complexity of climate change itself.

In my view, there is another dimension to our problem: the "climate" of climate change communication. The sense of climate change that has long dominated public discourse is a conception that truncates how danger is imagined, anticipated, and answered (at least in North America). This conception of climate results from the institutional commitment of governments to a particular "contact language" for interfacing between climate science and public life, what I call the "detection idiom," and it has encouraged the systematic under-prediction of dangers noted by Brysse et al. (2013). As I discuss below, the dominant conception of climate change is rooted in scientific methods designed to render it observable as global warming, a conception that limits discussions about what kinds of danger climate change poses and what kinds of responses we might implement. If we hope to process and discuss the dangers of climate change properly, we may first have to abandon our fascination with its detection.

The detection idiom

I consider the detection idiom a "contact language" (Galison 1997) that structures the relationship of climate science and public life through the use of an institutionally authorized symbol system. The idiom originated as a language suiting the needs of oceanographers and atmospheric scientists and expanded subsequently into a policy discourse for coordinating multiple actors across divergent political contexts. It is the interface that manages relationships between different social worlds by engendering action coordination around a pragmatic purpose.

The detection idiom is organized by the question of whether scientists are able to observe human influence in the global warming trend (what I call "the test"). The test has been a recurrent scientific fascination: first as a scientific curiosity; then as a vehicle for imagining how intentional weather modification might work; and today as a way of conceptualizing threats to valued resources and environments (Hart and Victor 1993). Dating initially to the nineteenth century, Gilbert Plass (1956a, 1956b) renewed scientific interest in the test by developing an approach for generating new evidence in the 1950s. He created a computer model for examining the effects of doubled carbon dioxide emissions (the indicator of human influence) on GMT (the indicator of climate change).

It is fair to say that significant investments would soon follow for the measurement of atmospheric carbon dioxide concentrations, the compilation of a global temperature record, and the modelling of "climate sensitivity" using global circulation models of the atmosphere. Oceanographers studying carbon cycles and atmospheric scientists researching the radiative balance of the planet used the tools of operational and mathematical meteorology (thermometer records and computer models) to develop a contact language for working together and to transform the question of human influence into scientific procedures able to "test" for detection (Weart 2001; Edwards 2010; see also Bohn 2011, for an important perspective on carbon dioxide measurement).

The detection idiom, at this point, is a scientific one, and it constitutes climate in terms of modern geophysics. While this idiom might have subsided in importance after detection of an anthropogenic signal in the warming record had been achieved, this shift did not happen. Instead, the idiom was entrenched further as a means of forwarding the goals of policy-framers who had already developed a regulatory approach informed by economic expertise that assumed the geophysicalist conception of climate. The policy imperatives established for the economic regulation of climate change would dominate how climate change was conceived and understood publicly.

The test, in brief, became what some scholars have labelled "the plan" (see Hulme 2013; Sarewitz 2011). The "plan" presumes that scientific demonstration of human influence on the global warming trend will facilitate the political consensus necessary for policy developments able to avert dangerous human interference with the climate (see Hulme 2013). The cultural authority of geophysics would suffice, it was hoped, as a driver of policy, one enabling global political organizations to gain traction on the matter.

In the detection idiom, trend detection (of global warming) and attribution (of human influence) are regarded as indispensable. The scientific understanding of carbon dioxide illustrates how emissions accumulate in the atmosphere, how this is converted to radiative forcing, and how that affects GMT. This way of approaching the question of global warming established the key terms and organizing logic for international policy:

> define a level of DAI in the climate response then, assuming knowledge of climate sensitivity, trace out what level of concentrations of CO_2 will get the world to that point, and then, assuming one understood the details of the carbon cycle, work back into what emissions scenarios will get the world to that concentration.
>
> *(Boykoff et al. 2010: 58)*

The detection idiom that had originated to coordinate different scientific actors in developing the test did not simply influence the prevailing approach to global policy. It also fixed the expectations for public discourse by prioritizing clear, certain, and consensus-based communication to facilitate policy and political change. Detection of anthropogenic climate change, first claimed in 1988 by James Hansen in his insistence that scientists must "stop waffling" on the matter, has been ceaselessly communicated ever since. The IPCC would subsequently claim detection with a wider array of evidence in 1995 (see Weart 2001). A steady flow of other techniques for demonstrating a scientific consensus on detection followed, including the recent claim that 97.2 per cent of scientists believe detection has been accomplished (see Cook et al. 2013). Communicative strategies designed to increase public uptake of this consensus on detection have also followed (van der Linden et al. 2014).

The detection idiom, in short, moved from its origins as a scientific test into policy deliberations and became the prevailing language that structures how climate science is situated within and circulates in public life, particularly in North America. Our problem is not the test, which was answered brilliantly by geoscientists and is widely deserving of praise and historical recognition. Our problem is that the public idiom derived from the test is insufficient for conceptualizing how danger is imagined, anticipated, and known, which is to say that the discussion of danger cannot be so neatly circumscribed as to permit scientists to determine when climate change is dangerous.

To explain: our dominant conception of climate change reflects its origins in US energy security concerns, and this legacy includes the deference that was (rightly) accorded to geophysical scientists in addressing the question of whether carbon dioxide emissions from human activity could influence climate ("the test"). By reducing climate change to what was called the "carbon dioxide question," geophysicists developed a contact language and assumed authority in providing a clear, certain, and uncontentious answer.

The respect for geophysics reflects the commitments to "high modernism" discussed by Scott (1999; see also Glover, Chapter 2, this volume), yet its public authority was generated through the ability to offer clarity, certainty, and consensus in its claims for climate change. The test could be answered by geophysics on these terms – something the social sciences and humanities could not do (see Hulme 2011b).

What of those dimensions of climate change that engender ambiguity, uncertainty, and diversity of opinion? A public language that is structured firmly by the expectations or demand for clarity, certainty, and consensus will distort the communication of these more intractable aspects. It is a demand that Stephen Schneider (1988) once called the "double ethical bind" of climate change communication. The norms of scientific practice were in tension with the norms of media coverage organized by the detection idiom.

The clarity, certainty, and consensus on detection encourage us to speak of dangers in a similar tone, or, failing that, to wait until the uncertain becomes more certain. If geoscientists can detect human influence on climate, one might reason, should they not also be able to detect when it becomes dangerous? This assumption also encourages policy-makers to conceptualize future risk in terms of metrics and methodologies that are more familiar, well understood, and culturally authoritative (Hulme 2011b). It is in this sense that the authority of geophysics in defining climate change as a matter of scientific detection (which, I believe, is correct) has led to its hegemony in shaping questions of risk and danger as matters of climate impact (which, I believe, is insufficient) – a process Hulme calls "epistemological slippage." Yet, and this is an important point, only a narrow slice of geophysical science has informed this public idiom, a point to which I will return in the conclusion.

The best indicator of this problem is the repurposing of the "global means" used to detect a human influence on the climate, carbon dioxide, and GMT as "global thresholds" for determining danger. Is it not odd that danger is imagined primarily in this way, using the very indicators developed to detect a human influence? While reasonable people can debate whether an increase of two degrees Celsius in GMT over pre-industrial levels is a wise policy goal to pursue at the global level, does it make sense to think of climate change danger in this singular way?

Hansen (2005) and Mahony and Hulme (2012) have argued that this approach has long foreclosed the discussion of danger by utilizing the metrics (GMT), reasoning (probability), and abstractions ("reasons for concern") encouraged by the detection idiom as institutionalized by the IPCC. Moreover, if Brysse et al. (2013) are correct in discerning a systematic bias towards under-prediction of dramatic or dangerous changes in the earth sciences, the cultural authority of such perspectives is a significant threat to the status quo. It is especially concerning that few, if any, formal discussions of danger at the Conference of the Parties were convened by the UNFCCC (Oppenheimer and Petsonk 2005), and that efforts to organize the diverse opinions of scientists and people on the subject have long been absent (Schneider 2001; Dessai et al. 2004).

To clarify: one might surmise, based on what I have written so far, that geophysicists have sought to gain sole authority in determining the future of climate change policy. However, this is not the case. As Samuel Randalls (2010, 2011) has shown, the GMT targets for determining dangerous thresholds reflect the way in which scientific analyses of climate change developed historically through policy contexts shaped by economists. In particular, the two-degree-Celsius target for avoiding danger reflects the dominance of the modelling of climate sensitivity in terms of the carbon dioxide and GMT relationship (what I have called "the test"), *as well as* the desire to understand likelihood, probability, and cost–benefit implications of particular scenarios ("the plan"). Randalls and his colleagues have also shown how this approach encouraged the long-term avoidance of danger – not to encourage adaptation or attention to ecological resilience (Boykoff et al. 2010). An architecture for avoiding danger will provide very weak analytical approaches for dwelling in danger.

It is tempting to say the development of the test (converting carbon dioxide emissions to carbon dioxide concentrations in the atmosphere to carbon dioxide radiative forcing and its effects on GMT) was simply inverted by the plan (set GMT threshold, find carbon dioxide concentrations limit, regulate carbon dioxide emissions). It would seem this approach was particularly attractive to economists seeking to use cost–benefit analyses to frame policy options:

> it is clear that the rationale for this target was generated in part from a conjoined economic–scientific philosophy. There is rather less scientific than economic merit in setting a temperature target to be used as the guiding policy for cost-effectively dealing with climate change.
> *(Randalls 2011: 240)*

If one accepts the argument as I have sketched it, and if one hopes to have a less constrained discussion regarding how climate change might become dangerous, and if one wishes to draw on the wider possibilities of geophysics (not simply the test for detection), social science (not simply economics), and humanistic work, how might one begin?

While many have proceeded critically by questioning the two-degree-Celsius target (Boykoff et al. 2010; Lenton 2011; Hulme 2013), I suggest revisiting the historical origins of an *ecological* conception of climate change, one that illustrates the narrowness – and I hope lessens the influence – of the "detection idiom."

It is this older ecological perspective, I suggest, that is reactivated in the newer warnings of climate change danger, and that can help us process the concern with extreme events and disaster. It is recognized in these warnings that the communicative strategy, policy framework, and institutional organization of the IPCC/UNFCCC construe climate change too narrowly, and are too beholden to the very political and economic arrangements impelling us towards danger. The dimensions of climate change that are inherently uncertain, that are the least tractable to generating consensus, and that cannot be avoided are the aspects of the situation that now demand our collective attention.

Gore's question of how to react to scientific warnings remains important. If we are to answer it, we must better understand how climate change is reframed as a multidimensional risk management problem, and we must ask if retrograde conceptions of risk and security will animate this reframing, or whether it will give way to a fuller and more reflexive appraisal of climate change.

Ecology and climate change: an alternative idiom

If the "detection idiom" is insufficient for matters of danger and risk management, as I suggest, its failures were not unanticipated. For instance, in 1985, W.C. Clark reviewed the past decade of governmental testimony delivered to US policymakers on "the carbon dioxide question" and concluded there was a deep disconnect between the research strategies of scientists and the knowledge desired by politicians. As governmental hearings and formal syntheses of science displaced (or disciplined) the "extreme event" style of public warnings, a detection idiom was congealing around global averages, likely scenarios, and demands for cost–benefit analyses of carbon dioxide regulations to frame the policy options.

It is worth recalling for a moment how frequently the public received warnings of existential danger in the 1960s and 1970s. These alerts resulted in part from the development of nationwide meteorological warning systems, but they were due mainly to the popularity of scientists speaking in unconstrained tones; Rachel Carson, Reid Bryson, Stephen Schneider, Sherwood Rowland, Maurice Ewing, Wally Broecker, Carl Sagan, and others, taken collectively, represent an "extreme event" conceptualization of environmental change that would fail to shape mainstream climate change discussion in the 1980s and 1990s.

Why? The US energy crisis, not the environmental movement, initiated political concern with climate change, and while Spencer Weart (2001), a historian of climate science, is certainly correct to suggest that both scientific and societal views co-evolved with respect to climate change as a result of environmentalism, it is also true that the central elements of climate change discourse have remained surprisingly resilient since the 1980s (see Boykoff et al. 2010; Randalls 2010).

The insecurities generated by US reliance on foreign sources of fossil fuel prompted an energy security debate in the late 1970s. Climate change, in this context, was primarily a question of the consequences of carbon dioxide emissions and a comprehensive study of the relationship of carbon dioxide to climate was requested through the Energy Security Act of 1980. By circumscribing the concerns of energy policy to carbon dioxide, the broader environmental challenge to industrial transformations of the atmosphere were bracketed out, even though governmental hearings at the time continually involved acid rain and stratospheric ozone depletion as shared concerns. Indeed, as Oreskes and

Conway (2010) have shown, it is impossible to understand the nature of climate scepticism without realizing how political opposition to government regulation developed through earlier efforts to undermine science on acid rain and stratospheric ozone. However, a circumscribed carbon dioxide issue could be reduced to a measureable question, one informed by geophysical scientists (not eco-freaks) and with a clear pathway for policy development, should it prove a pressing matter (carbon dioxide regulation). The economics modelling of energy policies, in particular, could then inform any consideration of impact or risk.

The National Academy of Sciences met the request for guidance by forming the Carbon Dioxide Assessment Committee and tasked it with providing the first comprehensive overview – both physical and social – of the implications of carbon dioxide emissions. In previous years, the physics community convened by the defence community and meteorological community respectively had filed reports on carbon dioxide as a potential policy problem (see MacDonald et al. 1979; NAS 1979). The 1983 NAS report *Changing Climate* differed from previous reports in two respects: it solicited and included more diverse expertise on climate change; and it sought to shape policy options as opposed to simply encouraging the involvement of policy-makers in climate change discussions.

Changing Climate has produced divisive historical assessments over the years because it opened a "policy window" for different sources of influence to compete over the direction of climate change policy. Some scholars suggest it interrupted "the emerging scientific consensus" (Oreskes et al. 2008: 6) by introducing "the kernel of the emerging skeptics' argument, and the eventual basis for the Reagan administration to push the problem off the political agenda entirely" (Oreskes and Conway 2010: 176). Others claim it simply reflected the difficult task of incorporating diverse expertise for the first time in a single document and thus reflected – rather than challenged – the slow evolution towards consensus (Nierenberg et al. 2010: 349). Spencer Weart (2001: 146) describes the report similarly – as hesitant and cautious rather than sceptical – yet does observe that it differed from an EPA (1983) report on carbon dioxide released days later: "The science was mostly the same, but the tone of the EPA's conclusions was more anxious."

The question of "tone" reflects the difficulties in moving from a consensus on detection of human influence on climate to a policy framework suitable for encompassing the risks and dangers of climate change. The hope is that an appropriate tone is the prerequisite for facilitating policy development. Yet, the tone appropriate to detection (clear, certain, and confident) is much more difficult to adopt for the question of climate impacts, as these are mediated by a much wider range of uncertainties than the relationship of carbon dioxide to GMT.

It was this problem that surfaced in the divergent reports reflected on by Weart. The science was not mostly the same, as Weart suggests, only the physical science. *Changing Climate* had taken the additional step of including social scientific and foreign policy concerns, and sought to integrate these insights into a comprehensive assessment. While some viewed the involvement of social sciences as necessary for policy development, and as part of an evolving approach, the way in which physical and social scientific concern are integrated has remained contentious to this day (Hulme 2013).

According to Oreskes and Conway (2010), the inclusion of social sciences permitted scepticism and climate denial to gain traction in authoritative policy discussions. Social scientists, Thomas Schelling in particular, expanded the conception of climate change beyond the question of carbon dioxide, emphasized the uncertainties in existing knowledge that resulted from this enlarged scope, and recommended regulatory delay until a wider range of societal options had been studied. These social scientists were economists, as Oreskes and Conway emphasize, and the hegemony of economics in speaking for societal concerns would prove an enduring problem (Hulme 2011a; Randalls 2011).

Schelling, however, is a difficult figure to characterize. His experience was hardly limited to the usual concerns of economics, and his approach was likely informed by the interactional approach he had developed to advise foreign policy strategists on arms control. I will return to this point shortly. First, however, I should emphasize that Oreskes and Conway are correct in their broader claim: the effect of Schelling's perspective was to muddy the emergent scientific consensus by embedding climate change within a broader range of societal and geopolitical dynamics, the future of which was unknown yet also constitutive of how the effects of carbon dioxide emissions would be distributed geographically and understood.

The carbon dioxide question would have to become the climate change problem on Schelling's way of thinking (see Randalls 2011). The question of the effects of climate change should not reflect the terms of the carbon dioxide question, as developed by geophysicists seeking to detect its influence on temperature trends, but expand to encompass a more interactional conception of the climate–society relationship, a point that was not unique to Schelling (see Bach 1980, cited in Randalls 2010). A policy framework commensurate with the issue would have to wait until this broader perspective was incorporated into policy analyses. As Randalls (2010) notes, *Changing Climate* was steering the discussion towards risk management, and its broadened conception of the problem was consistent with similar work published in Germany by Wilfrid Bach, particularly his co-edited collection *Man's Impact on Climate* (Bach et al. 1979). (Unfortunately, the very interesting developments in Germany remain largely outside the scope of this chapter.)

Oreskes and Conway (2010) point out that Schelling's views were challenging to incorporate into a synthesis, and *Changing Climate* remained more a collection of viewpoints rather than an integrated perspective. This is certainly the case. It becomes a problem, however, given Oreskes and Conway's conclusion that the summary and conclusions of *Changing Climate* were written to reflect Schelling's view, not the range of opinion in the report, and this was done by the chairperson of the NAS committee in order to incline a politically conservative framing for the document.

While these claims have generated some dispute, Oreskes and Conway (2010) are right to emphasize the tension between the social scientific and geophysical accounts of climate change. Schelling's work *did* encourage diverse sources of influence to compete over the direction of climate change policy and likely proved important to the salience subsequently given to economics. It is certainly possible that this window first permitted sceptical voices a toehold within influential governmental processes.

Schelling's work, however, was adopted not simply by those seeking to entrench economic concerns at the centre of climate policy, but also by ecologists advocating a "risk management" idiom for climate change. These ecologists were also challenging the incipient consensus. However, unlike economists, they were concerned with the narrowing of climate change to carbon dioxide given its exclusion of the kind of "extreme events" possibilities that now figure prominently in contemporary accounts of "the Anthropocene." It is this ecological approach to climate change and modelling that has resurfaced in the new warnings of danger over the last decade. Notably, it is this work that begins to incorporate a conception of "tipping points" into climate change science, an idea first developed through a dynamic systems approach by Schelling (Russill and Lavin 2012).

Schelling's contribution to climate change policy – and the risk management idiom that would develop from it – is best understood with respect to the integrated science, policy, and management approach developed at the International Institute for Applied Systems Analysis (IIASA) for the resilience of ecological systems. If geophysics would inform the dominant conceptualization of climate change to emerge from energy security concerns in the United States, it was systems ecology that grounded development of an alternative approach, one that brought together science, policy, and management to understand risk in terms of ecological resilience.

In the early 1980s, the ecological systems perspective developed by C.S. Holling, the director of IIASA, was extended to considerations of global climate change, primarily through work coordinated by W.C. Clark (1985), whose policy proposal to the 1985 Villach Conference (the event usually credited with establishing the direction of the IPCC) used Schelling's framework in *Changing Climate* to bring the broad concerns of IIASA to bear on "the carbon dioxide question." Paul Crutzen was another important influence on this approach, as I discuss below.

Clark (1985) credited Schelling extensively in the document, yet the influence clearly ran in both directions. Schelling (1996: 19) has mentioned Clark as an important influence in his own understanding of climate change policy and notes, in particular, that IIASA at the time was the only institute developing "integrated work on the subject." By "integrated work," Schelling means a particular approach to modelling. Hulme (2011b), for example, has pointed out that the clear distinction between climate impacts and climate–society interactions as alternative approaches to computer modelling had only just started to emerge in the 1980s, yet Schelling was endorsing an interactional understanding of climate change influenced by ecological modelling traditions.

Historical work on IIASA is badly needed. Schelling and William Nordhaus, the two social scientists involved in *Changing Climate*, both developed their initial views on climate policy in concert with IIASA (Nordhaus published the first economic model for climate change as a working paper for the institute in the 1970s; IIASA 2006). While Randalls (2011) offers a good account of the origins of Nordhaus' work, a similar account of Schelling in the context of IIASA would be useful. Absent this research, I will focus attention on how Clark's (1985) Villach proposal brought together the "extreme events" concerns of scientists with the dynamic systems modelling advanced by ecologists to reconfigure the carbon dioxide question as a multidimensional "risk management" problem.

There were several threads to Clark's analysis.

First, the carbon dioxide question was a "mess" (Clark 1985: 2) and needed to be treated as such. It was a mess because it was embedded within and mediated by social and ecological contexts at multiple scales and it could not be detached from any of those contexts. Yet, conceptual development was impeded by policy discussions that took "their questions from the immediate concerns of natural scientists studying carbon dioxide" (Clark 1985: 2). One could not make climate policy tractable in this way. One needed first to bring an integrated understanding of society, ecology, and climate into the conceptualization of climate change – not add policy implications to the framework established by scientists for the carbon dioxide question. The ensuing complexity would recommend a complexity science approach, one embraced by the dynamic systems modelling pioneered by Holling, Clark, and other ecologists (including Robert May).

Second, messes necessitated a different kind of policy approach: "In such a real world mess of multiple actors and actions, no-one's needs will be served by single 'bottom line' assessments that purport to speak for all people and all times" (Clark 1985: 2). If one sought this kind of singular and global assessment, it would cut off "the broader perspective that could help to locate the carbon dioxide question within the context of related economic trends, political agendas, and environmental problems" (Clark 1985: 3). The ecology of resilience, in particular, was quite sensitive to the way in which contextual features would mediate the impacts of global trends. One needed policies connected to goals that reflected the multiple and differently situated perspectives of diverse actors, as opposed to a global perspective that sought to transcend such matters to offer singular indictors or targets. The reductive idiom gaining authority for climate change – an idiom framed by natural scientists doing science – would encourage a reductive assessment process, in Clark's estimation.

Third, the environmental mess approach required the policy analysis community to view its task "as one of risk assessment and management" (Clark 1985: 36). In fact, Clark (1985: 36) argued that "the greatest single addition to usable knowledge about the carbon dioxide question might well come from recasting it as a problem of risk assessment and management." The main reason for Clark's conclusion here was the uncertainty that pervaded the subject. It would be necessary to manage and act on uncertainty – and we should not reduce uncertainty by directing scientific studies to those areas most likely to produce clarity and certainty. If one focused on averages, likelihood, and the cost–benefits of probable scenarios, one would summarily exclude abrupt changes and extreme events. Yet, these are the ways in which disruptions are experienced. Moreover, the lack of a developed risk management idiom opens the treatment of uncertainty "to unconstrained use as propaganda by all extremes of the political spectrum" (Clark 1985: 36). It was necessary to accept uncertainty and encompass it within a risk management approach – not reduce or ignore it. Modelling, after all, might well generate increased uncertainty, not eliminate it.

Fourth, one needed a more imaginative and interdisciplinary programme of policy exercises for gaining experience in projecting the future, one that could involve the collaborative production of "future histories" of how climate and society interact (Clark 1985: 46). Clark recognized that this process could disclose and force reflection on one's implicit assumptions, a feature that IIASA, in particular, had recently come to appreciate. Indeed, an internal dispute over energy modelling at IIASA, one that was grounded in different conceptions of how models reflected the exigencies of policy frameworks and their institutional contexts, had broken out in the years previous to Clark's (1985) proposal, and forced a more reflexive process of evaluation to the fore, one documented at length in Brian Wynne's (1984) account of the affair. Scenarios were thought experiments, not forecasts, and modelling could clarify and aid subjective judgement, not eliminate it. The contact language developed for imagining the future should recognize these features, not deprive public discourse of the pragmatic, flexible, craft-like nature of risk assessments. A policy language that fails to develop and accommodate "the *potential* for measured skepticism or qualified support … tends instead to polarize into all-or-nothing positions, and thus to deskill policy" (Wynne 1984: 314; emphasis in original). Clark, Wynne, and others would go on to recommend a "social learning" approach for these matters. Crutzen, of course, would subsequently welcome us to the Anthropocene.

The Clark proposal would not define the future of climate change policy. Geophysics, not ecology, and probabilistic concerns with metrics that changed gradually or cumulatively over time (like carbon dioxide concentrations or GMT) would be emphasized over extreme events or abrupt change possibilities. This foreclosing of a wider sense of climate change was bound up with a desire to make the matter tractable for states comprising the UN system in the 1990s.

It does seem that Clark's work encouraged a more gentle drift towards the incorporation of risk management approaches with which perspicuous historical work has rightly identified *Changing Climate* (Boykoff et al. 2010; Randalls 2010). While this risk management discussion has typically been submerged by the detection idiom institutionalized by the IPCC/UNFCCC processes, it would force the question of how to select for alternative futures in projecting impacts to the forefront of conversation (see Schneider 2001).

The legacy of the Clark proposal, however, lies in the alternative to the detection idiom that has developed with the widespread concern over tipping-point warnings of climate danger. Dating to 2005, Clark, Schellnhuber, and Crutzen would call for a "new paradigm," one bringing together Schellnhuber's claims for "earth system analysis" and Crutzen's identification of the Anthropocene (Clark et al. 2005). Moreover, the scientific formalization of tipping points would be published in a manuscript handled by Clark as managing editor of the *Proceedings of the National Academy of Sciences*. This new "risk management idiom," one inclusive of extreme events thinking, would compete to reconfigure climate change policy (see Lenton and Schellnhuber 2007), and the interest in tipping points would be led by systems-oriented ecologists in particular (Scheffer 2009).

The ecological politics of climate change had re-emerged, not in a "pure" form, of course, but in collaboration with a different approach to geophysics, one conceived as "earth system science." This "climate" for communication, a sense of climate change embracing both its socio-cultural constituents and more extreme, abrupt change possibilities in their effects on regions, has dislodged the physicalist conception of climate entrenched by the detection idiom, and permitted a wider sense of risk and danger to animate public discourse in recent years. On this account, dangers emerge, are mediated, and are addressed in a variety of ways, most of which are only loosely connected to changes in the GMT.

Conclusion

Have we answered Gore's question?

If there are proliferating warnings of danger, there are also myriad analogues for considering climate change risk (war, migration, public health, environmental precaution, ecological resiliency, extreme weather, emergency preparedness, military threat assessment, reinsurance industry actuarial analysis, risk society) as well as a burgeoning discourse on climate risk management, both public and within proprietary assessments of threats and vulnerability (for military, security, and financial institutions). The risk management idiom recommended by Clark's ecological and pluralist perspective would seem to complicate – rather than clarify – Gore's question of how we should react.

And it does.

Hansen, interestingly, broke definitively from the detection idiom in launching his iconic tipping-point warning. He pointed to radiative forcing more generally (not simply GMT), and sought to integrate a wider array of human emissions into his policy assessments (not simply carbon dioxide) in order to suggest the myriad ways in which the additional heat retained by the earth system might produce risk, including the notion now discussed as the Anthropocene – namely, the idea that the planet has entered a new epoch. Hansen's efforts permitted the broader concerns of geophysics – not simply the question of detection – to shape how warnings are developed, even if this effort pushed him well beyond his well-demarcated area of expertise (and into reflections on ice-sheet melting and hurricane intensity, for example).

The hurricane experts obliquely referenced in Gore's film also sought to broaden the discussion of danger beyond questions of anthropogenic detection of climate change. Hansen and Gore, like many others, were intensely frustrated by nefarious forms of industry-funded scepticism regarding climate change in 2006; and, in their discussion of Hurricane Katrina, it was the efforts to distort public discourse that framed the context in which Gore's film amplified the association of climate change and hurricane intensity. Post-Katrina, however, hurricane experts have argued that detection of an anthropogenic climate change signal in hurricane activity is not the primary issue. Framed in this way, the discussion of danger substitutes techniques of trend detection for the broad range of scientific and cultural techniques appropriate to risk analyses of disaster. For these scientists, it was the state approach to risk that was the primary problem for hurricane dangers, and it was in the context of a broader assessment of regional vulnerability that climate change should be considered (Emanuel et al. 2006). One needed to address the impoverished approach to risk embedded in the regulatory system and in the manufactured ecology, at least in the United States, if one wanted to put the question of climate change and hurricane disaster in its proper policy context (Emanuel et al. 2006). These arguments, I would suggest, reflect the "mess" approach to climate change advocated by Clark in the mid-1980s.

The demand for detection and attribution of a warning trend in the conditions of hurricanes reflects the entrenched nature of the "detection idiom," as I have described it, and it polarizes the discussion of uncertainty into conflictual claims that the evidence supports or fails to support the attribution of disaster to climate change (see Curry et al. 2006). It also calls us back to a conception of climate change rooted in the conventional concerns of energy security, back to the ways of conceptualizing and securing human life as biological processes distinct from broader geographical, ecological, and geophysical processes, and this, as Dalby (2013: 190) notes, is to foreclose rather than open up how we anticipate danger.

The better approach – the way to hear Gore's question – is to use the diverse analogues for risk management to develop a public idiom that accepts the uncertainty that can result from proliferating the indicators and measures of climate change danger. In terms of ecological geopolitics we should encourage the politics of regional vulnerability that would shape the development of this contact language. If this cannot be done, if one cannot transcend the failures of the detection idiom by revisiting how the relationship of the earth sciences and civic life can reflect commitments to the democratization of perspectives within political institutions, then the strategic imperatives of military planning, national security, and finance capitalism will dominate how risk and resiliency are conceptualized and acted upon. Indeed, there are good indications that this is quickly becoming the case, and this is a significant danger in itself.

References

Bach, W. (1980) "The carbon dioxide issue: What are the realistic options?," *Climate Change*, 3(1): 3–5.

Bach, W., Pankrath, J., and Kellogg, W. (eds) (1979) *Man's Impact on Climate*, Amsterdam: Elsevier.

Beck, U. (2010) "Climate for change, or how to create a green modernity?," *Theory, Culture & Society*, 27(2–3): 254–66.

Bohn, M. (2011) "Concentrating on CO_2: The Scandinavian and Arctic measurements," *Osiris*, 26(1): 165–79.

Boykoff, M.T., Frame, D., and Randalls, R. (2010) "Discursive stability meets climate instability: A critical exploration of the concept of 'climate stabilization' in contemporary climate policy," *Global Environmental Change*, 20: 53–64.

Brysse, K., Oreskes, N., O'Reilly, J., and Oppenheimer, M. (2013) "Climate change prediction: Erring on the side of least drama," *Global Environmental Change*, 23(1): 327–37.

Chakrabarty, D. (2009) "The climate of history: Four theses," *Critical Inquiry*, 35(2): 197–222.

Clark, W.C. (1985) "On the practical implications of the carbon dioxide question," IIASA Working Paper WP-85-043, prepared for UNEP in support of WMO/ICUS/UNEP meeting, Villach, Austria, 9–15 October.

Clark, W.C., Crutzen, P.J., and Schellnhuber, H.J. (2005) "Science for global sustainability: Toward a new paradigm," CID Working Paper No. 120, Cambridge, MA: Science, Environment and Development Group, Center for International Development, Harvard University (also published in H.J. Schellnhuber, P.J. Crutzen, W.C. Clark, M. Claussen, and H. Held (eds) *Earth System Analysis for Sustainability*, Cambridge, MA: MIT Press, pp. 1–28.

Cook, J., Nuccitelli, D., Green, S.A., Richardson, M., Winkler, B., Painting, R., Way, R., Jacobs, P., and Skuce, A. (2013) "Quantifying the consensus on anthropogenic global warming in the scientific literature," *Experimental Research Letters*, 8(2), http://iopscience.iop.org/1748-9326/8/2/024024/article, accessed 19 March 2015.

Curry, J., Webster, P.J., and Holland, G.J. (2006) "Mixing politics and science in testing the hypothesis that greenhouse warming is causing a global increase in hurricane intensity," *Bulletin of the American Meteorological Society*, 87(8): 1025–37.

Dalby, S. (2013) "Biopolitics and climate security in the Anthropocene," *Geoforum*, 49: 184–92.

Dessai, S., Adger, W.N., Hulme, M., Turnpenny, J., Köhler, J., and Warren, R. (2004) "Defining and experiencing dangerous climate change: An editorial essay," *Climatic Change*, 64: 11–25.

Edwards, P. (2010) *A Vast Machine*, Cambridge, MA: MIT Press.

Emanuel, K., Anthes, R., Curry, J., Elsner, J., Holland, G., Klotzbach, P., Knutson, T., Landsea, C., Mayfield, M., and Webster, P. (2006) "Statement on the US hurricane problem," 25 July, http://wind.mit.edu/~emanuel/Hurricane_threat.htm, accessed 9 January 2015.

Environmental Protection Agency (EPA) (1983) *Can We Delay a Greenhouse Warming? The Effectiveness and Feasibility of Options to Slow a Build-up of Carbon Dioxide in the*

Atmosphere, Washington, DC: Environmental Protection Agency, http://nepis.epa.gov/Exe/ZyPURL.cgi?Dockey=9101HEAX.txt, accessed 11 January 2015.

Galison, P. (1997) *Image and Logic: A Material Culture of Microphysics*, Chicago, IL: University of Chicago Press.

Guggenheim, D. (director) (2006) *An Inconvenient Truth: A Global Warning*. DVD. Beverly Hills, CA: Participant Productions.

Hansen, J. (1988) "Testimony before US Senate Committee on Energy and Commerce," Energy Policy Implications of Global Warming: Hearing before the Subcommittee on Energy and Natural Resources, 100th Congress, 1st session, 23 June.

Hansen, J. (2005) "Is there still time to avoid 'dangerous anthropogenic interference' with global climate? A tribute to Charles David Keeling," paper presented at the American Geophysical Union, San Francisco, CA, 6 December, www.columbia.edu/~jeh1/2005/Keeling_20051206.pdf, accessed 9 January 2015.

Hansen, J. (2006a) "Is there still time to avoid 'dangerous anthropogenic interference' with global climate?," paper presented at New School University, New York City, 10 February.

Hansen, J. (2006b) "The threat to the planet," *New York Review of Books*, 53(12), www.nybooks.com/articles/archives/2006/jul/13/the-threat-to-the-planet/?page=1, accessed 10 January 2015.

Hart, D.M. and Victor, D.G. (1993) "Scientific elites and the making of US policy for climate change research, 1957–1974," *Social Studies of Science*, 23(4): 643–80.

Hulme, M. (2011a) "Meet the humanities," *Nature Climate Change*, 1: 177–79.

Hulme, M. (2011b) "Reducing the future to climate: A story of climate determinism and reductionalism," *Osiris*, 26(1): 245–66.

Hulme, M. (2013) *Exploring Climate Change through Science and in Society*, New York: Routledge.

Illich, I. (1989) "The shadow our future throws," *New Perspectives Quarterly*, 26, www.digitalnpq.org/archive/2009_fall_2010_winter/12_illich.html, accessed 9 December 2014.

International Institute of Applied Systems Analysis (IIASA) (2006) "40 years' research into climate change," www.iiasa.ac.at/web/home/about/achievments/scientificachievements andpolicyimpact/From-Ice-Age-to-Heat-Wave.en.html, accessed 23 September 2014.

Lenton, T. (2011) "Beyond 2°C: Redefining dangerous climate change for physical systems," *WIREs Climate Change*, 2: 451–61.

Lenton, T.M. and Schellnhuber, H.J. (2007) "Tipping the scales," *Nature Reports*, 1, 97–8, www.nature.com/climate/2007/0712/pdf/climate.2007.65.pdf, accessed 9 January 2015.

McCarthy, J.J., Canziani, O.F., Leary, N.A., Dokken, D.J., and White, K.S. (eds) (2001) *Climate Change 2001: Impacts, Adaptation, and Vulnerability: Contribution of Working Group II to the Third Assessment Report of the Intergovernmental Panel on Climate Change*, Cambridge and New York: Cambridge University Press.

MacDonald, G.F., Abarbanel, H., Carruthers, P., Chamberlain, J., Foley, H., Munk, W., Nierenberg, W., Rothaus, O., Ruderman, M., Vesecky, J., and Zachariasen, F. (1979) *The Long Term Impact of Atmospheric Carbon Dioxide on Climate*, JASON Technical Report JSR-78-07, Arlington, VA: SRI International.

Mahony, M. and Hulme, M. (2012) "The colour of risk: An exploration of the IPCC's 'burning embers' diagram," *Spontaneous Generations*, 6(1): 75–89.

National Academy of Sciences (NAS) (1979) *Carbon Dioxide and Climate: A Scientific Assessment* [Charney Report], Washington, DC: National Academy of Sciences.

National Academy of Sciences (NAS) (1983) *Changing Climate*, Washington, DC: National Academy Press.

Nierenberg, N., Tschinkel, W.R., and Tschinkel, V.J. (2010) "Early climate change consensus at the National Academy: The origins and making of *Changing Climate*," *Historical Studies in the Natural Sciences*, 40(3): 318–49.

O'Neill, S. and Nicholson-Cole, S. (2009) "'Fear won't do it': Promoting positive engagement with climate change through visual and iconic representations," *Science Communication*, 30(3): 355–79.

Oppenheimer, M. and Petsonk, A. (2005) "Article 2 of the UNFCCC: Historical origins, recent interpretations," *Climatic Change*, 73(3): 195–226.

Oreskes, N. and Conway, E. (2010) *Merchants of Doubt: How a Handful of Scientists Obscure the Truth on Issues from Tobacco Smoke to Global Warming*, New York: Bloomsbury Press.

Oreskes, N., Conway, E., and Shindell, M. (2008) "From Chicken Little to Dr Pangloss: William Nierenberg, global warming, and the social deconstruction of scientific knowledge," *Historical Studies in the Natural Sciences*, 38(1): 109–52.

Plass, G.N. (1956a) "Effect of carbon dioxide variations on climate," *American Journal of Physics*, 24: 376–87.

Plass, G.N. (1956b) "The carbon dioxide theory of climate change," *Tellus*, 8(2): 140–54.

Randalls, S. (2010) "History of the 2°C climate target," *WIREs Climate Change*, 1: 598–605.

Randalls, S. (2011) "Optimal climate change: Economics and climate science policy (from heuristic to normative)," *Osiris*, 26: 224–42.

Russill, C. and Lavin, C. (2012) "Tipping point discourse in dangerous times," *Canadian Review of American Studies*, 42(2): 142–63.

Sarewitz, D. (2011) "Does climate change knowledge really matter?," *WIREs Climate Change*, 2(4): 475–81.

Scheffer, M. (2009) *Critical Transitions in Nature and Society*, Princeton, NJ: Princeton University Press.

Schelling, T. (1996) "Research by accident," *Technological Forecasting and Social Change*, 53(1): 15–20.

Schneider, B. and Nocke, T. (2014) *Image Politics of Climate Change: Visualizations, Imaginations, Documentations*, New York: Columbia University Press.

Schneider, S.H. (1988) "The greenhouse effect and the US summer of 1988: Cause and effect or a media event: An editorial," *Climatic Change*, 13(2): 113–15.

Schneider, S.H. (2001) "What is 'dangerous' climate change?," *Nature*, 411: 17–19.

Schneider, S.H. (2006) "Laurie David + Stephen Schneider: The activist and climate scientist enter the Seed Salon to deliberate the state of the planet," *Seed Magazine*, 24 April, http://seedmagazine.com/content/article/laurie_david_stephen_schneider/, accessed 9 January 2015.

Scott, J.C. (1999) *Seeing Like a State: How Certain Schemes to Improve the Human Condition Have Failed*, New Haven, CT: Yale University Press.

van der Linden, S.L., Leiserowitz, A.A., Feinberg, G.D., and Maibach, E.W. (2014) "How to communicate the scientific consensus on climate change: Plain facts, pie charts or metaphors?," *Climatic Change Letters*, 126(1–2): 255–62.

Weart, S. (2001) *The Discovery of Global Warming*, Cambridge, MA: Harvard University Press.

Wynne, B. (1984) "The institutional context of science, models and policy: The IIASA Energy Study," *Policy Sciences*, 17(3): 277–320.

4
DISCONNECTING CLIMATE CHANGE FROM CONFLICT

A methodological proposal

Emily Meierding

In April 2014, World Bank President Jim Yong Kim asserted that "Fights over water and food are going to be the most significant direct impacts of climate change in the next five to ten years. There's just no question about it' (quoted in Elliott 2014). Such dire predictions have become commonplace among policy-makers. They are also articulated by climate change mitigation advocates who attempt to use the threat of climate conflicts to mobilize political action. A number of widely reported academic studies have also identified linkages between climate change and intra-state armed conflict. Yet most academic research on climate conflicts has produced ambiguous or negative results. Contrary to statements like Kim's, there is no robust evidence that climate change is linked, directly or indirectly, to violent contention.

Most scholars have responded to these analytical ambiguities by calling for more climate conflict research. Underpinning this proposal is a belief that further methodological refinements, aimed at bringing statistical models closer to theorized climate conflict connections, will produce more consistent and reliable empirical findings. This assumption has some merit; the methods employed in quantitative climate conflict research have improved significantly over the last few years and additional advances could enhance our understanding of the connections between climate change and intra-state conflict. However, maintaining the current research agenda also comes at a cost. Basic methodological fixes will not address the normative problems that arise from researchers' current framing of climate conflicts. Consequently, in this chapter, I argue for a more radical reframing of climate conflict research. Specifically, I encourage scholars to disconnect climate change and conflict analyses either by removing climate change from conflict studies or by studying the full range of social responses to climate change, without privileging conflict.

This proposal is prompted by two normative concerns. One is that recent methodological adjustments threaten to erase broader responsibilities for climate change and climate change-related conflicts. By shifting the geographical scale of their analyses to the sub-national level and focusing on the proximate drivers of civil contention, researchers are producing more compelling conflict models. However, in making these adjustments, they move further away from climate change. As a result, the blame for climate conflicts is relocated. Violence is attributed to developing countries' local scarcities rather than developed countries' overconsumption of energy and food resources. This approach frames climate conflicts as "their" problem, rather than "our" fault, and removes developed countries' moral imperative to engage in climate conflict prevention.

Second, current climate conflict research perpetuates popular beliefs that violence is the modal response to negative environmental change, especially in less developed countries. Even if scholars find little evidence of climate conflict linkages, by framing their studies around armed conflict, they erase alternative social responses to climate change. In so doing, they misrepresent the range of strategies that people employ to manage environmental degradation and scarcity, and minimize the agency and creativity of local populations. These tendencies are exacerbated by biased journalistic coverage of climate conflict research. Studies that find a statistically significant correlation between climate change and conflict tend to receive widespread coverage, while critiques of these analyses and studies that find no relationship are ignored. As a result, academic climate conflict research reinforces problematic environmental conflict framings even if most scholars are, themselves, producing unbiased work.

The argument is presented in two sections. First, the chapter assesses quantitative climate conflict analyses, noting four shortcomings of early models: problems with independent variable data; dependent variable data; the scale of analysis; and a focus on direct, universal causal effects. This section also identifies methodological fixes that have become increasingly widespread in quantitative climate conflict research, including the adoption of new datasets, geographically disaggregating analyses, and considering intervening variables and conditional effects.

The second section highlights the unintended negative consequences of these methodological developments and argues that, to mitigate them, scholars should reframe their analyses, removing climate change from conflict studies or examining a broader set of social responses to climate change. This section also considers the rhetorical consequences of disconnecting climate conflict analyses, concluding that the proposed reframing will not impede climate change mitigation efforts.

The hunt for climate conflict connections

Popular and academic interest in climate conflicts took off in the mid-2000s. In 2003, a report written for the US Department of Defense described how "abrupt climate change" could lead to violent contention (Schwartz and Randall 2003). In 2007 and 2008, a number of policy reports identified gradual climate change as a threat to national and human security (CNA Corporation 2007; US National Intelligence Council 2008; WBGU 2008). In addition, the connections between climate change and security were debated in the United Nations Security Council, where many delegates asserted that climate change-induced shortages of water and food could exacerbate instability and lead to violent contention, especially in "fragile states" (UNSC 2007).

At the same time, academic researchers began to take up the question of whether climate change could inspire intra-state armed conflict. A special issue of *Political Geography* was devoted to the topic in 2007. Early climate conflict research built on two existing literatures: one on environmental security and one on civil wars and armed conflicts. Theories were drawn largely from the former; like environmental security research, climate conflict research emphasizes the impact of environmental degradation and resource scarcity. Methodologically, however, climate conflict research has drawn more from the latter field. Rather than relying on case studies, as did most environmental security analyses (Baechler 1998; Homer-Dixon and Blitt 1998), many climate conflict scholars adopted the quantitative methods prevalent in civil war studies, especially multivariate statistical analysis, to assess the linkages between climate change and intra-state armed conflict. This chapter focuses on these quantitative analyses, as their findings have received the greatest popular attention.

Many early climate conflict studies essentially incorporated climatological variables into existing civil war models. These climate conflict models evaluated the impact of hydrometeorological disasters, or short-term deviations in temperature and precipitation, on the frequency of intra-state armed conflict while controlling for other factors that are commonly found to encourage civil conflict, such as economic performance and population size (Hendrix and Glaser 2007; Nel and Righarts 2008; Burke et al. 2009). Other studies tested the direct impact of temperature and precipitation on conflict without employing socio-economic or political controls (Zhang et al. 2007; Tol and Wagner 2010).

These early analyses produced mixed results. A widely reported study by Burke et al. (2009) identified a connection between warmer temperatures and armed conflict; the authors went so far as to specify the number of deaths that would occur as a result of global warming over the following two decades. However, other scholars challenged the robustness of this finding, observing that it was highly dependent on model specification (Buhaug 2010). Most other early climate conflict studies produced similarly inconsistent results. Within individual studies, some climatological variables were connected to conflict, while others were not. The impact of specific variables also differed across studies. Consequently, assessments of the early literature concluded that relationships between climate change and conflict were far from certain (Salehyan 2008).

There are two possible explanations for authors' failure to identify consistent connections between climate change and intra-state armed conflict. One is that no such relationship exists and climate conflict models simply revealed that absence. The other explanation is model misspecification. The models that were tested in early climate conflict research deviated significantly from theorized linkages between climate change and intra-state contention. Most researchers expect that, if climate conflicts occur, they are predominantly localized, intercommunal events (Buhaug et al. 2008; Gleditsch 2012). Contentious episodes do not necessarily involve the central state or the entirety of a country. The impacts of climate change are also mediated by social and political institutions; shifting climatological conditions do not automatically cause violent contention under all circumstances.

Yet the statistical models tested in early climate conflict research often failed to capture these features. They deviated from theorized climate conflict connections in at least four ways. First, studies were conducted at the national level rather than the local level. Second, to operationalize "conflict," the dependent variable, researchers employed datasets that systematically excluded small-scale and non-state contention. Third, for their explanatory variables, researchers examined short-term changes in weather rather than long-term changes in climate. Fourth, analyses rarely considered intervening variables or conditional effects. As a result, it is unsurprising that models failed to produce consistent findings. And it is reasonable to assume, as most researchers did, that the reasons for this failure were model misspecification rather than the absence of climate conflict connections. The remainder of this section elaborates on each of these problems, as well as on the steps quantitative researchers have taken to improve their model designs, bringing theory and empirical analyses closer together.

Analysing climate conflicts at the national level is problematic because climatological and hydrometeorological conditions can vary widely within a country; one region may experience normal conditions, while another suffers from a severe drought. This variation is even more pronounced in the case of hydrometeorological disasters, such as floods, hurricanes, and typhoons – events that are expected to become more frequent and more severe as a result of climate change. Not all geographical areas will be affected by these disasters. Similarly, conflict events rarely encompass an entire country. Even civil wars can leave significant portions of states untouched. Smaller-scale conflicts are likely to be restricted to a single region, city, or even village. Given this geographic localization, it is quite possible for a hydrometeorological deviation to occur in one part of a country, while an unrelated contentious episode occurs in another. A national-level study would mistakenly assume that these events were connected. It would inaccurately identify the contentious event as a climate conflict, biasing results.

To respond to this problem, many researchers have shifted the scale of their analyses. Some have collected data at the level of administrative provinces (Meier et al. 2007; Fjelde and von Uexkull 2012). Others have divided areas under study into 50 km^2, 100 km^2, or 1° grids, collecting data on climate and conflict within each square (O'Loughlin et al. 2012; Raleigh and Kniveton 2012; Theisen 2012; Wischnath and Buhaug 2014a). By spatially disaggregating their studies, researchers reduce the risk of false positives; it is more likely that climate events and conflict events that occur within a smaller geographical area are related. However, researchers may not have access to similarly fine-grained data on political and socio-economic factors, so some theoretical mismatch persists.

In addition, many researchers have continued to conduct analyses at the national level (Hendrix and Salehyan 2012; Koubi et al. 2012; Slettebak 2012).

The second methodological problem with early climate conflict analyses was their use of civil war and armed conflict datasets. This reliance was unsurprising given many researchers' backgrounds in civil war studies and the challenges of collecting new conflict data. However, these datasets are not well suited to analysing climate conflicts as they systematically exclude episodes of localized, small-scale, and communal contention. The most commonly employed dataset is PRIO/Uppsala's collection of armed conflicts (Gleditsch et al. 2002). This dataset includes all contests in which the state was a participant and that resulted in at least 25 battle deaths. Any contests that did not meet the battle-death threshold, or did not involve the central government or its representatives, are excluded. Therefore, all studies that use this data overlook the smaller-scale, non-state contests that are widely expected to be the more frequent consequences of climate change (Hendrix and Glaser 2007; Nel and Righarts 2008; Tol and Wagner 2010; Hsiang et al. 2011; Theisen et al. 2011/12; Bergholt and Lujala 2012; Koubi et al. 2012; Slettebak 2012; Wischnath and Buhaug 2014a). One study, by Burke et al. (2009), exacerbated the mismatch by examining only civil war conflicts that had resulted in at least a thousand battle deaths.

To correct for these omissions, a few researchers have recently employed alternative conflict data. These alternative datasets include a wider range of actors or a wider range of contentious events. Fjelde and von Uexkull (2012) use the Uppsala Conflict Data Program (UCDP) Non-State Conflict Dataset, which retains the 25 battle-death threshold but includes conflicts that did not involve the central state (Sundberg et al. 2012). Hendrix and Salehyan's (2012) Social Conflict in Africa Dataset (SCAD) includes a wider range of events: coups, military infighting, violent repression by the government, extra-governmental violence, peaceful protests, riots, strikes, mutinies, and communal conflicts, as well as conventional armed conflicts. Raleigh and Kniveton (2012) and O'Loughlin et al. (2012) use the Armed Conflict Location and Event Dataset (ACLED), which disaggregates armed conflicts into individual violent acts perpetrated by the government, rebels, militias, and anti-government rioters and protesters (Raleigh et al. 2010).

Other researchers have compiled micro-level datasets of conflict patterns in individual countries or regions. Theisen's (2012) study of climate conflicts in Kenya uses local news sources to identify acts of inter-group violence and state-perpetrated violence within the country. An early study by Meier et al. (2007) uses field monitor reports from the Inter-Governmental Authority on Development's Conflict Early Warning and Response Network (CEWARN) to collect data on armed clashes, raids, protests, and banditry in Ethiopia, Kenya, and Uganda. The results of such sub-national studies may not be generalizable beyond the included countries; however, by providing a closer match between theory and data, their specific findings are more compelling.

Climate change is a broad phenomenon. Its anticipated physical consequences include shifts in temperature and rainfall patterns, such as the timing and location of monsoon seasons, accelerated sea level rise, habitat changes for animal and plant species, and increased frequency and intensity of hydrometeorological disasters, such as storms, droughts, floods, landslides, extreme temperatures, and wildfires. These changes could be gradual and linear or they could be large and abrupt, occurring after a climate system reaches a tipping point. This potential for unexpected, dislocating shifts is one characteristic of climate change that distinguishes it from the resource scarcities that were analysed in earlier environmental security research, such as shortages of freshwater and arable land (Meierding 2013). Two other distinguishing factors of climate change are the breadth of its effects on the earth's physical systems and its temporal scope; climatological changes are commonly measured over decades or longer.

Climate conflict research has not incorporated many of these characteristics. Most climate conflict studies have examined short-term changes in weather rather than long-term changes in climate. They operationalize climate change as month-to-month or year-to-year changes in temperature and precipitation, or as monthly or annual deviations from long-term means. Alternatively, a number of studies have examined the impact of hydrometeorological disasters on conflict (Nel and Righarts 2008; Bergholt and Lujala 2012; Slettebak 2012). Yet, while natural disasters and shifts in temperature and precipitation are anticipated consequences of climate change, these studies remain one step removed from the broader concept. They would more accurately be described as "weather conflict" or "natural disaster conflict" research than "climate conflict" research. Other anticipated consequences of climate change have also received little attention in conflict studies. Scholars have not evaluated the impact of accelerated sea level rise, habitat changes, or shifting monsoon patterns on intra-state contention. Nor have the impacts of abrupt, non-linear climatological shifts been assessed.

Some researchers have attempted to study the effects of multi-century adjustments in weather and precipitation patterns. However, these studies face sizable methodological trade-offs (Zhang et al. 2007; Tol and Wagner 2010). In order to expand models' temporal scope, data must be collected at a continental or global scale. The only available conflict data for these extended time frames are counts of inter-state wars. Data on socio-economic and political conditions are also unavailable. To avoid these trade-offs, Hsiang et al. (2011) employed an alternative measure of climatological shifts: the El Niño Southern Oscillation (ENSO). Thus, the study maintained the country–year unit of analysis that is common in climate conflict research, while using a uniquely climatological explanatory variable. However, such innovations have been rare. Most researchers, recognizing the difficulty of conducting temporally and geographically disaggregated analyses using climatological explanatory variables, have continued to employ short-term weather variation and hydrometeorological disasters as their measures of climate change.

Contrary to the World Bank President's claim that conflict will be a direct effect of climate change, most commentators and analysts recognize that causal pathways from climate change to conflict pass through multiple intervening variables. If climate change inspires conflict, this occurs through its effects on factors such as agricultural productivity, freshwater availability, and migration. Yet researchers' models often fail to include these intervening factors. Instead, many analysts assess the direct impacts of climate measures on conflict (Hendrix and Glaser 2007; Meier et al. 2007; Nel and Righarts 2008; Burke et al. 2009; Tol and Wagner 2010; Hsiang et al. 2011; Theisen et al. 2011/12; Fjelde and von Uexkull 2012; Hendrix and Salehyan 2012; O'Loughlin et al. 2012; Raleigh and Kniveton 2012; Slettebak 2012; Theisen 2012).

Some researchers have attempted to capture more of the causal pathway from climate change to conflict by employing two-stage least-squared (2SLS) models. This analytic technique evaluates the effect of climate measures on conflict via an intervening variable, such as economic growth. The strategy was initially employed as a methodological fix by researchers who wanted to assess the impact of economic growth on conflict but were concerned about endogeneity problems. They observed that while economic growth could influence conflict, conflict could also influence economic growth. In order to overcome this problem, researchers used precipitation and temperature measures – two truly exogenous variables – as instruments for economic growth (Miguel et al. 2004). Later, this method was adopted by researchers who were explicitly interested in the effects of precipitation and temperature changes via intervening variables (Bergholt and Lujala 2012; Koubi et al. 2012; Wischnath and Buhaug 2014a). The causal pathway from climate change to conflict, through economic growth, has therefore received some attention. However, other intermediate variables, such as agricultural productivity or migration, have not been employed in two-stage models.

Instead, these other variables have received some attention in single-stage analyses. Zhang et al. (2007) tested the impact of agricultural prices and agricultural yield on the frequency of conflict in China and Europe over extended time frames. Arezki and Brückner (2011) assessed the effects of high food prices, possibly brought on by climate-related agricultural downturns, on the likelihood of contention. Wischnath and Buhaug (2014b) examined the effect of changing levels of wheat and rice production on conflict in India. Urdal (2008) also conducted a sub-national study of conflict in India, assessing the impact of agricultural yields on contention. Other studies have examined the impact of water scarcity on internal conflict (Gizelis and Wooden 2010; Böhmelt et al. 2014). In addition, Buhaug and Urdal (2013) tested the impact of urban growth, which could be prompted by climate change, on public unrest. Notably, most of these analyses do not frame themselves solely as climate conflict studies. However, the authors allude to climate change, noting that it could cause the environmental and social shifts that are the focus of their research.

Another way that researchers have attempted to refine their tests of climate conflict connections is to consider conditional effects. Climate change has commonly been described as a "threat multiplier" (CNA Corporation 2007; see Dalby, Chapter 6, this volume, for further discussion). Its impacts are not expected to be equal in all geographic areas or for all people. Regions and groups that are already physically and socially vulnerable are expected to suffer more from its effects (Raleigh 2010). Vulnerable populations are also expected to be more likely to turn to violence in response to climate change. To incorporate these conditional effects, some researchers have added interaction terms to their models. They test whether the impact of climate measures on conflict depend on factors such as poverty, political access, ethnic marginalization, population density, agricultural productivity, and levels of democracy (Theisen et al. 2011/12; Fjelde and von Uexkull 2012; Koubi et al. 2012; O'Loughlin et al. 2012; Theisen 2012; Wischnath and Buhaug 2014a). Sometimes, these analytical efforts are rather cursory. Nevertheless, by attempting to specify further the conditions in which climate conflicts are most likely to occur, these modifications improve our understandings of the causal processes that could lead from climate change to contention.

Researchers have made progress towards bringing climate conflict theories and models closer together. A significant mismatch still exists between climate change, conceptualized broadly, and the particular hydrometeorological measures that predominate in climate conflict analyses. However, most recent studies have employed at least one of the other modifications: shifting the geographical scale of analysis; employing more appropriate conflict data; or specifying causal pathways via intermediate variables and conditional effects. Some models include all three of these adjustments. Yet, recent reviews of climate conflict literature still conclude that cumulative positive findings within the field have been limited. Most studies continue to reveal few strong connections between climate change and violent contention. Moreover, when climate measures are found to have an impact, their influence on conflict often pales in comparison to economic and political variables (Gleditsch 2012; Scheffran et al. 2012; Meierding 2013; Theisen et al. 2013; Kallis and Zografos 2014). Reviewers have nonetheless resisted rejecting the climate conflict hypothesis. Instead, they recommend further research, incorporating the methodological modifications identified here.

Reframing climate conflict analyses

This chapter, in contrast, argues that scholars should pursue a more radical reframing of climate conflict research. By maintaining the current research agenda, quantitative climate conflict researchers continue to frame armed violence as the modal response to environmental degradation and scarcity. This framing erases the alternative ways that individuals, groups, and governments respond to climate change. In addition, by continuing to frame their analyses as climate conflict research, while moving further away from climatological variables and shifting to sub-national levels of analysis, scholars inadvertently transfer blame for climate conflicts from developed countries' consumption patterns to less developed countries' environmental scarcities. Together, these two dynamics create an image of climate conflicts as highly localized, almost deterministic events, occurring primarily in less developed countries. This portrayal is particularly influential because of inconsistencies in the popular reporting of climate conflict research. Studies that find climate conflict connections garner headlines; critiques and inconclusive analyses do not.

Climate conflict researchers' tendency to focus on violent contention is understandable, given the field's inheritances from civil war and environmental security studies. Conflict is also normatively important. Since violent contention can induce great human suffering, it is unsurprising that researchers want to understand its causes and dynamics in order to enhance prevention efforts. The ubiquity of the belief that climate change induces contention also encourages research; scholars have a professional interest in evaluating popular hypotheses. Moreover, conflict is clearly a political issue and thus an obvious fit with the disciplines of political science and international relations.

However, violence is only one possible social response to climate change. Individuals and groups employ a variety of coping strategies and adaptive measures to respond to shifting climatological conditions. They can collaborate to manage scarce resources, seek alternative employment, temporarily or permanently migrate, collect remittances or aid, change crops, obtain credit, consume savings or food stocks, create community networks and cooperatives, sell livestock, or forage (Ostrom 1990; Ellis 1998; Raleigh et al. 2008; Morrissey 2013). These alternative activities are likely to be more common responses to climate change than conflict. Violence is costly and dangerous. It is difficult to organize and retaliation can be severe. Consequently, people are likely to avoid it, especially when alternative coping and adaptive strategies are sufficient to sustain their lives and livelihoods.

Climate conflict scholars recognize that armed conflicts are rare events, while environmental changes, degradation, and scarcities are widespread. Much of their research has also found few connections between climate change and conflict. Yet, by continuing to frame their analyses around conflict, researchers sustain the impression that violence is the only active response to climatological shifts. In climate conflict models, people can either engage in conflict or do nothing; alternative actions are unspecified. The use of intervening variables makes the pathways from climatological shifts to conflict appear more conditional; these models demonstrate that violence is not an automatic response to climate change under all circumstances. However, as the models' only possible outcomes remain "conflict" or "no conflict," these adjustments do little to challenge the perception that violence is a common response to environmental change. Moreover, the models still imply that violence can occur, fairly deterministically, under *some* circumstances: in particular, when populations are poor or ethnically marginalized, or when political institutions are weak.

Only a few quantitative studies have included alternative responses to changing environmental conditions in their statistical models, and none of these has examined climate change per se. Böhmelt et al. (2014) assess whether freshwater scarcity inspires conflict or cooperation. Slettebak (2012) suggests that climate change-induced natural disasters might lead to cooperation. However, like most climate conflict studies, the latter's model evaluates only the occurrence of conflict.

As a result, a wide range of human responses to climate change is overlooked (Javeline 2014). These omitted activities are also those that cast populations in a more positive light. Behavioural adjustments require ingenuity and determination. By employing non-violent responses to climate change, individuals demonstrate their willingness to modify their activities and lifestyles in ways that do not violate established social codes. Many coping strategies and adaptive activities reveal people's capacity for cooperation and altruism rather than contention and self-interest. These positive qualities are erased when scholars exclude such actions from their analyses. Instead, people's responses to climate change seem quasi-animalistic; in response to scarcity, they either fight (Barnett 2000) or appear to do nothing. This depiction is particularly problematic because much climate conflict research focuses on less developed countries: in particular, states in sub-Saharan Africa. By focusing on conflict and eliminating the possibility of active, non-violent responses to climate change, climate conflict studies risk perpetuating images of savagery or passive victimhood in the face of environmental shifts. To minimize these stereotypical portrayals, researchers should reframe their analyses to examine a wider range of social responses to climate change.

The significance of this reframing extends beyond academia. The topic of climate conflicts has attracted widespread interest among policy-makers and the general public. As a result, climate conflict research receives significant journalistic coverage. However, reporters tend to favour findings that reproduce conventional environmental conflict frames: in particular, the "resource war" narrative, which suggests that environmental degradation and scarcity lead directly to inter-group violence. As a result of this bias, studies that identify strong climate conflict connections attract widespread media coverage. Most recently, Hsiang et al.'s (2013) meta-analysis of quantitative climate conflict research caught the attention of the popular press. By combining over 50 climate conflict models, this study found that climate change had exacerbated many different types of contention. However, to reach this conclusion, the researchers violated a number of standard disciplinary methodological practices (Buhaug et al. 2014). In addition, they presented an exceptionally deterministic view of climate conflict connections by failing to include control variables (Raleigh et al. 2014). Nevertheless, the findings were widely reported, reinforcing popular perceptions that climate change is a direct cause of conflict.

It is unlikely that critiques or studies that fail to find a climate conflict connection will receive similar coverage. An earlier analysis by Burke et al. (2009) that identified a statistically significant connection between climate change and civil wars received similar media attention, in spite of its significant methodological flaws. Subsequent critical analyses, which strongly challenged Burke et al.'s findings, did not (Buhaug 2010). These inconsistent reporting practices perpetuate problematic assumptions about the prevailing social consequences of climate change, particularly in less developed countries. They also create an image of consensus in the popular media that does not exist in academic debates.

There are significant obstacles to incorporating alternative social responses into quantitative analyses. The largest are data-related: we do not have good measures of these alternative behaviours (Oppenheimer 2013). Non-violent responses to climate change are likely to be highly context specific and therefore not amenable to global data collection efforts. They are also hard to identify as they are less prominent than violent conflicts and garner far less journalistic attention. As a result, data on alternative responses to climate change would probably have to be gathered locally, through methods like interviews and household surveys. These approaches represent a significant deviation from current disciplinary practices. Such activities are also less overtly political than violent contention, which could discourage international relations and political science researchers. Nevertheless, there is potential scope for interdisciplinary collaboration between conflict scholars and anthropologists or development specialists.

If quantitative conflict researchers are incapable of, or not interested in, shifting the scope of their analyses to incorporate a broader range of social responses to climate change, they should consider dropping the topic of climate conflicts altogether. As was noted above, researchers' efforts to analyse the causes of conflict more accurately are leading their models ever further away from climate change. Almost no scholars test the impact of purely climatological phenomena on the likelihood of conflict. And some have eliminated hydrometeorological disaster and precipitation and temperature data from their analyses entirely, focusing instead on later variables in the causal chain, such as freshwater availability, agricultural yield, food prices, and urbanization (Zhang et al. 2007; Urdal 2008; Gizelis and Wooden 2010; Arezki and Brückner 2011; Buhaug and Urdal 2013; Böhmelt et al. 2014; Wischnath and Buhaug 2014b). Few of these studies are specifically climate conflict analyses. Yet, most still refer to climate change as a likely cause of the environmental or social shifts they are studying. This presentation implies that the contentious episodes they analyse are, to some degree, climate conflicts.

The danger of maintaining this linkage, while removing climatological variables and conducting geographically disaggregated analyses, is that awareness of the broader causes of climate conflicts will diminish. Climate change is not a purely local phenomenon. It is caused primarily by overconsumption of energy resources in developed states. Thus, if climate conflicts occur, the responsibility rests, to a large extent, outside of the countries experiencing contention. When climate change figures prominently in empirical models, this fact is harder to ignore. However, when studies are one or more steps removed from climate change and rely exclusively on geographically disaggregated data, the causes of conflict appear to be purely local: local environmental scarcities; local inter-group hostilities; and local institutional weaknesses. These modelling practices, while enhancing our understanding of conflict, erase broader culpability. This distancing is amplified by climate conflict researchers' focus on less developed states (Mayer 2012). By obscuring the connections between consumption patterns in more developed countries and conflict in less developed countries, these analyses create the impression that climate conflicts are "their" fault. "We" are not held accountable.

Methodologically, however, it is difficult for analysts to include climate variables in their models and conduct convincing tests of specific hypotheses about the causes of intra-state contention. Researchers face a trade-off: better understandings of conflict or continuing to highlight climate change. This chapter advocates a full break between the two fields. Scholars should conduct conflict studies *or* they should analyse the social effects of, and responses to, climate change. This shift would enable conflict researchers to sustain recent methodological improvements without erasing broader responsibilities for climate change.

This proposal is likely to prompt resistance, both from scholars who are still committed to evaluating the empirical relationships between climate change and conflict, and from climate change mitigation advocates who view climate conflicts as a useful rhetorical device. These advocates use the risk of climate conflicts to highlight the intensity of climate change-related suffering in less developed countries and to engender a sense of threat in developed countries. To accomplish the former, they suggest that climate change has caused or intensified violent intra-state contention in areas such as Darfur, Nigeria, and Syria. For the latter, they connect climate conflicts to population displacements, instability, and terrorism that might threaten developed countries. These strategies garner headlines, yet their efficacy in inspiring climate change mitigation efforts is unclear. Advocates have been linking climate change to armed conflict for at least a decade, yet international political action on climate change during that period has been limited (see Dalby, Chapter 6, this volume, for further discussion).

There are also alternative, less normatively problematic and more politically effective rhetorical strategies that advocates can deploy in order to inspire climate change action: in particular, emphasizing the broader negative consequences of climate change. Conflict is not the only possible negative impact of climate change. Hydrometeorological disasters destroy homes and businesses. Shifting weather patterns undermine livelihoods. Efforts to cope with these events strain national budgets. These consequences have an impact on both more and less developed countries. By speaking to widely held concerns (Sarewitz and Pielke 2000), this framing creates a sense of commonality and empathy rather than reinforcing the "us versus them" divide. In addition, these threats are more immediate and tangible to the populations of more developed countries than the supposed dangers of climate conflicts, so they are more likely to inspire political action. Hence, moving the focus away from climate conflicts should not harm climate change mitigation efforts. Meanwhile, there is much to gain from making the shift.

References

Arezki, R. and Brückner, M. (2011) *Food Prices and Political Instability*, International Monetary Fund Working Paper, WP/11/62, www.imf.org/external/pubs/ft/wp/2011/wp1162.pdf, accessed 10 June 2014.

Baechler, G. (1998) "Why environmental transformation causes violence: A synthesis," *Environmental Change and Security Project Report*, 4: 24–44.

Barnett, J. (2000) "Destabilizing the environment–conflict thesis," *Review of International Studies*, 26: 271–88.
Bergholt, D. and Lujala, P. (2012) "Climate-related natural disasters, economic growth, and armed civil conflict," *Journal of Peace Research*, 49: 147–62.
Böhmelt, T., Bernauer, T., Buhaug, H., Gleditsch, N.P., Tribaldos, T., and Wischnath, G. (2014) "Demand, supply, and restraint: Determinants of domestic water conflict and cooperation," *Global Environmental Change*, 29: 337–48.
Buhaug, H. (2010) "Climate not to blame for African civil wars," *Proceedings of the National Academy of Sciences*, 107: 16477–82.
Buhaug, H. and Urdal, H. (2013) "An urbanization bomb? Population growth and social disorder in cities," *Global Environmental Change*, 23: 1–10.
Buhaug, H., Gleditsch, N.P., and Theisen, O.M. (2008) *Implications of Climate Change for Armed Conflict*, Washington, DC: World Bank Group, Social Dimensions of Climate Change Workshop.
Buhaug, H., Nordkvelle, J., Bernauer, T., Böhmelt, T., Brzoska, M., Busby, J.W., Ciccone, A., Fjelde, H., Gartzke, E., Gleditsch, N.P., Goldstone, J.A., Hegre, H., Holtermann, H., Link, J.S.A., Link, P.M., Lujala, P., O'Loughlin, J., Raleigh, C., Scheffran, J., Schilling, J., Smith, T.G., Theisen, O.M., Tol, R.S.J., Urdal, H., and von Uexkull, N. (2014) "One effect to rule them all? A comment on climate and conflict," *Climatic Change*, 127: 391–7.
Burke, M.B., Miguel, E., Satyanath, S., Dykema, J.A., and Lobell, D.B. (2009) "Warming increases the risk of civil war in Africa," *Proceedings of the National Academy of Sciences*, 106: 20670–4.
CNA Corporation (2007) *National Security and the Threat of Climate Change*, www.cna.org/reports/climate, accessed 10 June 2014.
Elliott, L. (2014) "Climate change will 'lead to battles for food,' says head of World Bank," *Guardian*, 3 April, www.theguardian.com/environment/2014/apr/03/climate-change-battle-food-head-world-bank, accessed 5 January 2015.
Ellis, F. (1998) "Household strategies and rural livelihood diversification," *Journal of Development Studies*, 35: 1–38.
Fjelde, H. and von Uexkull, N. (2012) "Climate triggers: Rainfall anomalies, vulnerability and communal conflict in sub-Saharan Africa," *Political Geography*, 31: 444–53.
German Advisory Council on Global Change (WBGU) (2008) *World in Transition: Climate Change as a Security Risk*, London: Earthscan.
Gizelis, T.I. and Wooden, A.E. (2010) "Water resources, institutions, and intrastate conflict," *Political Geography*, 29: 444–53.
Gleditsch, N.P. (2012) "Whither the weather? Climate change and conflict," *Journal of Peace Research*, 49: 3–9.
Gleditsch, N.P., Wallensteen, P., Eriksson, M., Sollenberg, M., and Strand, H. (2002) "Armed conflict 1946–2001: A new dataset," *Journal of Peace Research*, 39: 615–37.
Hendrix, C.S. and Glaser, S.M. (2007) "Trends and triggers: Climate, climate change and civil conflict in sub-Saharan Africa," *Political Geography*, 26: 695–715.
Hendrix, C.S. and Salehyan, I. (2012) "Climate change, rainfall, and social conflict in Africa," *Journal of Peace Research*, 49: 35–50.
Homer-Dixon, T.F. and Blitt, J. (eds) (1998) *Ecoviolence: Links among Environment, Population, and Security*, Lanham, MD: Rowman & Littlefield.
Hsiang, S.M., Burke, M., and Miguel, E. (2013) "Quantifying the influence of climate on human conflict," *Science*, 341(6151), www.sciencemag.org/content/341/6151/1235367.full, accessed 20 March 2015.

Hsiang, S.M., Meng, K.C., and Cane, M.A. (2011) "Civil conflicts are associated with the global climate," *Nature*, 476: 438–41.

Javeline, D. (2014) "The most important topic political scientists are not studying: Adapting to climate change," *Perspectives on Politics*, 12: 420–34.

Kallis, G. and Zografos, C. (2014) "Hydro-climatic change, conflict and security," *Climatic Change*, 123: 69–82.

Koubi, V., Bernauer, T., Kalbhenn, A., and Spilker, G. (2012) "Climate variability, economic growth, and civil conflict," *Journal of Peace Research*, 49: 113–27.

Mayer, M. (2012) "Chaotic climate change and security," *International Political Sociology*, 6: 165–85.

Meier, P., Bond, D., and Bond, J. (2007) "Environmental influences on pastoral conflict in the Horn of Africa," *Political Geography*, 25: 716–35.

Meierding, E. (2013) "Climate change and conflict: Avoiding small talk about the weather," *International Studies Review*, 15: 185–203.

Miguel, E., Satyanath, S., and Sergenti, E. (2004) "Economic shocks and civil conflict: An instrumental variables approach," *Journal of Political Economy*, 112: 725–53.

Morrissey, J.W. (2013) "Understanding the relationship between environmental change and migration: The development of an effects framework based on the case of northern Ethiopia," *Global Environmental Change*, 23: 1501–10.

Nel, P. and Righarts, M. (2008) "Natural disasters and the risk of violent civil conflict," *International Studies Quarterly*, 52: 159–85.

O'Loughlin, J., Witmera, F.D.W., Linke, A.M., Laing, A., Gettelman, A., and Dudhia, J. (2012) "Climate variability and conflict risk in East Africa, 1990–2009," *Proceedings of the National Academy of Sciences*, 109: 18344–9.

Oppenheimer, M. (2013) "Climate change impacts: Accounting for the human response," *Climatic Change*, 117: 439–49.

Ostrom, E. (1990) *Governing the Commons: The Evolution of Institutions for Collective Action*, Cambridge: Cambridge University Press.

Raleigh, C. (2010) "Political marginalization, climate change, and conflict in African Sahel states," *International Studies Review*, 12: 69–89.

Raleigh, C. and Kniveton, D. (2012) "Come rain or shine: An analysis of conflict and climate variability in East Africa," *Journal of Peace Research*, 49: 51–64.

Raleigh, C., Jordan, L., and Salehyan, I. (2008) *Assessing the Impact of Climate Change on Migration and Conflict*, Washington, DC: World Bank Group, Social Dimensions of Climate Change.

Raleigh, C., Linke, A., and O'Loughlin, J. (2014) "Extreme temperatures and violence," *Nature Climate Change*, 4: 76–7.

Raleigh, C., Linke, A., Hegre, H., and Karlsen, J. (2010) "Introducing ACLED: An armed conflict location and event dataset special data feature," *Journal of Peace Research*, 47: 651–60.

Salehyan, I. (2008) "From climate change to conflict? No consensus yet," *Journal of Peace Research*, 45: 315–26.

Sarewitz, D. and Pielke, R. Jr. (2000) "Breaking the global-warming gridlock," *Atlantic Monthly*, 86: 54–64.

Scheffran, J., Brzoska, M., Kominek, J., Link, P.M., and Schilling, J. (2012) "Climate change and violent conflict," *Science*, 336: 869–71.

Schwartz, P. and Randall, D. (2003) *An Abrupt Climate Change Scenario and Its Implications for United States National Security*, report commissioned by the US Department of Defense, http://eesc.columbia.edu/courses/v1003/readings/Pentagon.pdf, accessed 10 June 2014.

Slettebak, R.T. (2012) "Don't blame the weather! Climate-related natural disasters and civil conflict," *Journal of Peace Research*, 49: 163–76.

Sundberg, R., Eck, K., and Kreutz, J. (2012) "Introducing the UCDP non-state conflict dataset," *Journal of Peace Research*, 49: 351–62.

Theisen, O.M. (2012) "Climate clashes? Weather variability, land pressure, and organized violence in Kenya, 1989–2004," *Journal of Peace Research*, 49: 81–96.

Theisen, O.M., Gleditsch, N.P., and Buhaug, H. (2013) "Is climate change a driver of armed conflict?," *Climatic Change*, 117: 613–25.

Theisen, O.M., Holtermann, H., and Buhaug, H. (2011/12) "Climate wars? Assessing the claim that drought breeds conflict," *International Security*, 36: 79–106.

Tol, R.S.J. and Wagner, S. (2010) "Climate change and violent conflict in Europe over the last millennium," *Climatic Change*, 99: 65–79.

United Nations Security Council (UNSC) (2007) "United Nations Security Council holds first ever debate on impacts of climate change," www.un.org/News/Press/docs/2007/sc9000.doc.htm, accessed 10 June 2014.

Urdal, H. (2008) "Population, resources, and political violence: A subnational study of India, 1956–2002," *Journal of Conflict Resolution*, 52: 590–617.

US National Intelligence Council (2008) *Global Trends 2025: A Transformed World*, Washington, DC: US Government Printing Office, www.dni.gov/files/documents/Newsroom/Reports%20and%20Pubs/2025_Global_Trends_Final_Report.pdf, accessed 10 June 2014.

Wischnath, G. and Buhaug, H. (2014a) "On climate variability and civil war in Asia," *Climatic Change*, 122: 709–21.

Wischnath, G. and Buhaug, H. (2014b) "Rice and riots: On food production and conflict severity across India," *Political Geography*, 43: 6–15.

Zhang, D.D., Brecke, P., Lee, H.F., He, Y.Q., and Zhang, J. (2007) "Global climate change, war, and population decline in recent human history," *Proceedings of the National Academy of Sciences*, 104: 19214–19.

5
CLIMATE JUSTICE

Climate change, resource conflicts, and social justice

Paul Routledge

Social justice, conflict, and climate change

Resource scarcity increasingly defines the nature of twenty-first-century societies characterized by insecurities and conflicts, rooted in uneven distributions of resources. The results of these inequalities are manifest in livelihood vulnerability, poor health outcomes, and environmental degradation – all of which are exacerbated by the impacts of climate change (IPCC 2008). The context for such inequalities is the relentless expansion in global demand for natural resources as well as models of development predicated on resource extraction wherein Global Northern states and transnational corporations (TNCs) project economic, social, and political power into the Global South, resulting in resource "enclosures" (or "accumulation by dispossession"; Harvey 2003). Accumulation by dispossession has entailed the privatization of key "common" resources (such as public utilities and public institutions; genetic material) and state redistributions (such as the privatization of social housing, health, and education) as well as the deregulation of the global financial system (and associated speculation and corporate fraud) and the management and manipulation of financial crises (such as the debt trap and structural adjustment programmes) (Harvey 2003; Routledge and Cumbers 2009).

In the Global North, this has led to increased precariousness among vulnerable sectors of society, increased unemployment, austerity policies, and increased inequality, leading to reduced health and well-being (Wilkinson and Pickett 2009). In the Global South, there have been widespread dispossessions from poor, peasant, and indigenous peoples of vital resources, such as food and water, leading to health and livelihood crises. In both North and South, the cultures and practices of indigenous and small-scale food-producing communities are being steadily eroded, extinguishing their economic and social human rights while undermining the in situ genetic diversity and associated knowledge that will be necessary to respond to future environmental change (Barthel et al. 2013).

These kinds of processes have led to increasing conflicts over key land, food, water, and energy resources (Klare 2001; Le Billon 2004; Carmody 2009; see also Meierding's discussion, Chapter 4, this volume) and policy responses framed around discourses of security (see Dalby's discussion, Chapter 6, this volume) that frequently exacerbate resource dispossessions. Responses by communities and associated social movements to such approaches have given rise to discourses and practices of climate justice that express economic, political, and social *emancipatory* power across space and seek to imbue responses to climate change with a primary concern over social justice (Wright 2010; Levien 2013; Mountz 2013; Martinez-Alier et al. 2014).

In this chapter I argue that climate justice comprises three co-constitutive tendencies of antagonism, common(s), and solidarity (Chatterton et al. 2013). I will show how these tendencies play out in the politics of occupation conducted by the Bangladesh Kriskok Federation (Bangladesh Farmers' Federation; BKF) and the Bangladesh Kishani Sabha (Bangladesh Women Farmers' Association; BKS). I first started working with these movements in 2002 in my role as one of the facilitators of the People's Global Action (Asia) (PGA (Asia)) network in which the BKF participated (Routledge 2008). My research strategy has involved politically engaged and committed research that is practice-based and conducted in horizontal collaboration with social movements (Routledge 1996, 2002; Juris 2007, 2008). This has meant participating with the BKF and BKS in helping to organize solidarity-building activities such as an international conference that took place in Dhaka, Bangladesh, in 2004 (Routledge 2008).

First, I will briefly discuss the history and meaning of the term "climate justice" before discussing socio-environmental change in Bangladesh and responses from landless peasants.

Climate justice

The term "climate justice" was first used in a 1999 report that appeared on a website[1] followed by a November 2000 conference of the National Committee for Sustainable Development (NCDO) of the Netherlands, held in Amsterdam during the COP6 (Conference of the Parties) climate change negotiations. The concept gained further elaboration in: the Bali Principles of Climate Justice, 2002; the Durban Declaration on Carbon Trading, 2004, articulated by the Durban Group for Climate Justice; the formation of the Climate Justice Now! (CJN) network[2] in Bali, Indonesia, during the COP14 negotiations in 2007; and the Climate Justice Action (CJA) network as an organizing platform prior to the COP15 Copenhagen mobilizations. Subsequently, articulations of climate justice have been taken forward through: the World People's Conference on Climate Change in Bolivia, 2010; the mobilizations during the COP16 in Cancun, Mexico; and the COP17 in Durban, South Africa.[3]

Briefly defined, climate justice refers to principles of democratic accountability and participation, ecological sustainability, and social justice, and their combined ability to provide solutions to climate change. Such a notion focuses on the interrelationships between, and addresses the root causes of, the social injustice, ecological destruction, and economic domination perpetrated by the underlying logics of pro-growth capitalism. In particular, climate justice articulates a rejection of capitalist solutions to climate change (such as carbon markets) and foregrounds the uneven and persistent patterns of eco-imperialism and "ecological debt" as a result of the historical legacy of uneven use of fossil fuels and exploitation of raw materials, offshoring, and export of waste (see Muradian and Martinez-Alier 2001; Martinez-Alier 2002).

Building on the Climate Justice Now! declarations in 2007 and 2008, climate justice principles were articulated in the Klima Forum's declaration during the Copenhagen COP15 mobilizations. They included leaving fossil fuels in the ground; reasserting peoples' and communities' control over production; re-localizing food production; massively reducing overconsumption, particularly in the Global North; respecting indigenous and forest peoples' rights; and recognizing the ecological and climate debt owed to the peoples in the Global South by the societies of the Global North, necessitating making reparations. In a further elaboration, the Cochabamba Declaration of 2010 has argued for a series of "Inherent Rights of Mother Earth"[4] and demanded that developed countries radically reduce and absorb their emissions; assume the costs and technology transfer needs of developing countries and responsibility for climate refugees; eliminate their restrictive immigration policies, offering migrants a decent life with full human rights guarantees in their countries; and construct an adaptation fund to assess the impacts and costs of climate change in developing countries and provide a mechanism for compensation (Chatterton et al. 2013).

In short, such a rendition of climate justice involves a politics of antagonism, commons, and solidarity. In Bangladesh, the politics of land occupation "brings people into an antagonistic relation with capital," especially through "the active creation of different ways of organising existence" (Building Bridges Collective 2010: 83). In so doing, this positions occupation in relation to broader antagonisms around the uneven and exploitative social and environmental relations, as well as broader trajectories of contestation that attempt to make the power relations that comprise neoliberal capitalism localizable and contestable. Further, such antagonisms are generative of actions to create, defend, and expand the common(s), especially given accumulation by dispossession. Finally, occupation as a spatial and social practice is generative of new solidarities that extend beyond the local (Routledge 2003; Featherstone 2008; Juris 2008; Routledge and Cumbers 2009). Before grounding such claims in the realities of land struggle in Bangladesh, I will discuss socio-environmental change in that country.

Socio-environmental change in Bangladesh

Bangladesh is located in the "tropic of chaos," where the impacts of the catastrophic convergence of climate change, poverty, and violence are most acutely felt (Parenti 2011). It is considered to be one of the most vulnerable countries to climate change and sea level rise (IPCC 2008). Rising sea levels along its coast are already occurring at a greater than the global rate (of 1.0 to 2.0 mm/year) due to global sea level rise and local factors such as tectonic setting, sediment load, and subsidence of the Ganges delta (Karim and Mimura 2008). Further, the coastal region is particularly vulnerable to cyclonic storm surge floods due to its location in the path of tropical cyclones, the wide and shallow continental shelf, and the funnelling shape of the coast (Paul and Dutt 2010). Eighty per cent of the country consists of floodplains of the Ganges, Brahmaputra, Meghna, and other rivers, which sustain 75 per cent of the country's 160 million people (in 2011; Brouwer et al. 2007). The majority of the country's population are poor and dependent on agriculture, and are thus more vulnerable to the impacts of changing climatic regimes, particularly flooding (Dasgupta et al. 2011; Doyle and Chaturvedi 2011; Gilman et al. 2011).

The hegemonic production of nature is enacted through the liberalization of agrarian markets (for example, through the imposition of structural adjustment programmes), the expansion of biotechnologies in agriculture (for example, through the genetic modification of seeds and crops), and the privatization of nature through the extension of intellectual property rights to agriculture products (for example, through patent protection laws that enable agribusiness to convert scientific knowledge into commodities and charge fees for their use) (Nally 2011). Since the early 1990s, the government of Bangladesh has implemented structural adjustment programmes, including trade liberalization of agriculture, involving the withdrawal of input subsidies, privatization of fertilizer distribution and seed production, and elimination of rural rationing and price subsidies (Murshid n. d.). These have increased farmers' indebtedness and landlessness as they struggle to secure the capital to pay for expensive agricultural inputs (see also Desmarais 2007). Functional landlessness (i.e., ownership of less than 0.2 ha) affects 69 per cent of the population (Hossain 2009; Seabrook 2013). Brought about through land grabs by rural elites, local government corruption, and environmentally induced displacement, landlessness deterritorializes the poor.

Environmental risk exposure is increased for those with low incomes and less access to land (Brouwer et al. 2007). While the country's capacity to deal with cyclones has improved through the establishment of early cyclone warning and evacuation systems and cyclone shelters, leading to a decrease in fatalities, the capacity of existing cyclone shelters is woefully inadequate to accommodate all of the people in flood risk areas (Karim and Mimura 2008; Paul and Dutt 2010). Poor peasants' vulnerability is also exacerbated by hazard risk perceptions generated by influences of local culture, behaviour, and coping strategies, as well as inadequate land management policies and transport infrastructures (Chowdhury 2009; Alam and Collins 2010).[5]

Moreover, the Government of Bangladesh's Climate Change Strategy and Action Plan (BCCSAP), concerned with food security, adaptation, mitigation, and comprehensive disaster management, has been primarily shaped by bureaucrats, senior economists, non-government organizations (NGOs) such as the Equity and Justice Working Group Coalition and the Oxfam-led Campaign for Sustainable Rural Livelihood, and international donors such as the UK Department for International Development (DFID). Those most vulnerable to climate change – the rural poor – were largely absent from the plan's formulation and as yet little has been initiated in terms of policy (Ayers and Huq 2009; Raihan et al. 2010; Alam et al. 2011). Hence, for social movements such as the BKF and BKS, the challenges of climate change fold into ongoing conflicts over access to key socio-environmental resources, such as land. Such conflicts entail antagonisms (over the exclusion of poor peasants from access to land, and from climate change decision-making), commons (since collectively occupied land opens up issues of common resource use among peasants), and solidarity (since social movement struggles necessitate social relations of common ground among peasants). It is to these processes that I now turn.

Occupation as antagonism

Climate justice positions politics in relation to the unequal and contested geographies of power in terms of antagonism. This is in contrast to mainstream debates around climate change that have frequently isolated processes like carbon emission and global warming from the unequal social and environmental relations upon which neoliberal globalization depends (Giddens 2009) and have marginalized the responses of social movements and other grassroots initiatives (Panitch and Gindin 2011). Further, as Erik Swyngedouw (2007, 2010) has argued, climate change has been constructed as a "consensual" or "post-political" issue wherein the "consensus" on how to deal with it entails "rebooting" capitalism, creating new opportunities for accumulation, overcoming present failures, and increasing market penetration and resource/land privatization.

In contrast, climate justice actions around the world are politicizing climate change through making capitalist business-as-usual localizable and contestable: for example, ongoing protests against the exploitation of tar sands in Canada; protests in the UK, such as climate camps (located at sites of fossil fuel emissions such as the Drax and Kingsnorth power stations and Heathrow Airport[6]) and campaigns against new coal exploitation in Scotland and Wales;[7] and ongoing efforts to achieve a moratorium on coal and oil exploration in Nigeria and South Africa (Chatterton et al. 2013).

Climate justice activities consist of networked constituencies of activists who articulate an antagonistic politics of climate change beyond and below the nation state (as the alter-globalization protests did earlier; see Routledge 2003; Featherstone 2008; Routledge and Cumbers 2009). They also produce a set of political interventions that can be usefully described as "environmentalisms of the poor," which contest assumptions that environmental alliances and tactics are middle-class privileges (Martinez-Alier 2002; see also Martinez-Alier and Temper 2007; Featherstone 2008). They generate perspectives that are antithetical to further capital expansion and develop movements that want to tackle not just climate change but also the unequal social and environmental relations in which carbon emissions are embedded, and locate climate change within the broader crisis of contemporary capitalism. This brings us to the land struggle in Bangladesh.

The BKF was established in 1976, and the BKS in 1990. They are now estimated to have a total of 1.5 million members (interview, Dhaka, Bangladesh, 2011). Both social movements have established "social movement spaces" that are national in scope of operations (their joint office is located in Dhaka), while focused on specific, place-based occupations throughout Bangladesh. The BKF and BKS also participate in South Asian and other international networks.[8]

In 1987, the national government introduced a land law that enabled landless people to occupy and farm fallow (*khas*) land. However, the landless have faced continued government inaction on implementation of this law, so, since 1992, the BKF and BKS have organized landless people to occupy approximately 76,000 acres of *khas* land. Most of these occupations are concentrated in the south of the country (i.e., the part most vulnerable to climatic events) and land has been distributed to more than 107,000 of the poorest men and women living in the countryside (interview, Dhaka, Bangladesh, 2011).

Antagonism is articulated first against local and national government officials who have failed to implement the 1987 law and against large landowners and their private armies of armed thugs (*goondas*) who attempt to grab such land for themselves. One activist referred to the violence perpetrated by big landowners, their *goondas*, and the police who have made several attacks on the landless people's settlements: "Peasant activists are attacked, beaten, burned, jailed, and their homes are burned. That is the reality that we face" (interview, BKF activist, Kurigram District, Bangladesh, 2009). Through occupation, the BKF seeks to reconfigure peasant livelihoods in order to construct land controlled and farmed by peasants.

However, the place-specific conditions of local organizing act to constrain the capacity of the BKF and BKS to coordinate their struggles. According to activists, these include: first, the time taken up with the struggles to occupy land and the necessities of maintaining the occupation; second, a paucity of resources, particularly a lack of funds available to prosecute land occupations; and third, the impacts of (local and national) government actors and policies that favour corporate agri-business and local elites. As one activist argued:

> For a successful land occupation, the movement needs a strong occupation committee, whose leaders can withstand attacks by the landlords' *goondas*; a strong mass mobilization; a medical team who can provide medical treatment to those who suffer physical attacks; and a legal team to fight the

legal cases brought by landlords in the local courts in an attempt to stop the occupation. This takes time and money. Then we occupy the land. We build makeshift shelters for the occupying families, and provide food relief until the peasants can sow *padi* [rice]. The peasants must drink river water, and many get sick, until we have dug tube wells.

(Interview, Kurigram District, Bangladesh, 2009)

Such antagonisms are generative of actions to create, defend, and expand resource "commons," such as land and food production, to which I now turn.

Antagonism and commons: food sovereignty

The antagonism underpinning land occupation is not simply towards certain aspects of injustice; it also concerns how life is produced and reproduced and whether it is produced in common or not. The "common" refers to the social process of being-in-common, a social relationship of the commoners who build, defend, and reproduce the commons. The "commons" refers to territorial entities and those resources that are collectively owned or shared between and among populations as well as socio-nature – the air, water, soil, plants, and so on of nature, as well as the results of social (re)production and interaction, such as knowledge, languages, codes, and information (Hardt and Negri 2009; Building Bridges Collective 2010).

Dispossessions from poor, peasant, and indigenous peoples of vital resources and attacks on their livelihoods have generated moves to defend the common(s), which in turn generate further antagonisms against those class interests that seek to undermine them. Indeed, the planet is riven by struggles over resources and territory that evoke the dynamic generative process of commoning. Struggles such as the land occupations of the Movimento dos Trabalhadores Rurais Sem Terra (Movement of Landless Rural Workers; MST) in Brazil, the Zapatista Autonomous Municipalities of Chiapas, Mexico, the South African Shack Dwellers movement, and the BKF and BKS in Bangladesh are indicative of attempts to obtain social wealth and collectively organize social (re)production through antagonistic politics that directly challenge resource dispossessions of the poor (Wolford 2010; Routledge 2011; Chatterton et al. 2013).

In Bangladesh, antagonism takes the form of opposition to the attempted hegemony of market-led agriculture. The BKF and BKS fight what Antonio Gramsci (1971) termed a "war of position" (i.e., struggles waged within civil society that challenge the ideological and material basis of elite rule) organized around issues of food sovereignty, implying (farmer) control over territory, biodiversity, and the means of (food) production. In so doing, they both resist neoliberal constructions of common sense (i.e., the ideology of the primacy of market mechanisms and privatization of resources as the most efficient means of alleviating poverty and inequality) and practise alternative forms of production and conceptions of nature–society relations (Karriem 2009; Carroll 2010). Thus, occupation is then consolidated through practices of food production, socialization, and the reproduction of labour power and everyday life, which Federici (2010, 2012) terms a "reproductive commons."

The purpose is to integrate the political and material rights and practices of peasant farmers into the interactions and interdependencies between the social, economic, reproductive, and environmental dimensions of rural life (Wittman 2010), as noted by the BKF's president:

> In order to overcome the present challenges of the climate crisis, the small peasant can play a pivotal role ... Food sovereignty means the people's right to produce and consume culturally appropriate and accepted healthy and adequate food and their right to define their own food and agriculture policy ... It is not based on the need of TNCs. It prioritizes the local and national economy, peasants and family farm-based agriculture, artisan style fishing, pastoralist-led grazing and food production, distribution and consumption based on environmental, social and economic sustainability.
>
> *(Alam 2008: 4)*

While definitions of food sovereignty vary between organizations, and activist networks have changed over time and contain inconsistencies, common themes have emerged, such as: direct democratic participation and agrarian reform, implying peasant control over territory, biodiversity (commons), and means of (food) production; self-governance; ecological sustainability; the articulation of cultural difference; and so on. These have acted as a point of encounter, common interest, and solidarity between farmers' movements and international farmers' networks, such as La Via Campesina (The Peasant's Way; LVC), to which the BKF and BKS belong (Patel 2009; see also Rosset 2003; Windfuhr and Jonsen 2005; Holt-Gimenez and Patel 2009). It is claimed that such food sovereignty farming practices attempt to repair the dynamic and interdependent process that links society to nature through labour that has been undermined by the exploitation of socio-nature through capitalist agriculture (Wittman 2009), and enable peasant communities to mitigate and adapt to the effects of climate change, because of the biological resistance of crops, the recovery capacity of land, and the interdependent social dynamics between peasants (Desmarais 2007; La Via Campesina 2009; Altieri 2010; Rosset et al. 2011).

Through land occupation and membership of LVC, the BKF and BKS's struggle for the commons in Bangladesh is also viewed as a central demand/practice of translocal climate justice networks, rather than as something that is necessarily bounded or particular (Gilroy 2010; Routledge 2011). Occupation and practices of food sovereignty are prefigurative – they attempt to practise the future that they wish to see – and have the potential to generate solidarities (Franks 2003). Indeed, struggles to (re)create locally controlled commons, especially for the most marginalized, are also parts of broader struggles to mount connected geopolitical challenges to move the present balance of power away from ever more powerful coalitions of multinational institutions and to strengthen a globally connected grassroots movement for greater climate justice (Chatterton et al. 2013). It is to this final climate justice tendency that I now turn.

Climate justice solidarities

Solidarities, both within and between social movements, are forged out of the collective articulations of different place-based struggles, and constituted as the varied interconnections, relations, and practices between participants (Featherstone 2005). In particular, what requires negotiation is the politics of extension and translation of place-based interests and experiences (Katz 2001) in order for productive connections to be generated between different places and organizations (see, for example, Featherstone 2012).

Shared notions of climate justice begin to create common ground, enabling different themes to be interconnected, and different political actors from different struggles and cultural contexts to join together in common struggle (della Porta et al. 2006). While the national scale remains critical for the BKF for attempts to mobilize scarce resources, the need to construct more spatially extensive solidarity networks is also recognized:

> We prioritize direct action in trying to stop carbon emissions. Our strategy is to continue to struggle here against the government but also internationally to push our demands. We are fighting for alternatives, for land, but we need to build networks.
> *(Interview, Dhaka, Bangladesh, 2009)*

The BKF and BKS participate in a variety of networks that developed in response to the threats posed to peasant livelihoods by neoliberal globalization, including the Aaht Sangathan (Eight Organizations) in Bangladesh, whose total number of members is now close to two million (interviews, Dhaka, Bangladesh, 2004, 2009), LVC, the Asia Peasants Coalition, the People's Coalition on Food Sovereignty, and the Social Forum processes. Concerning climate justice, the BKF, together with LVC, the Asia Peasants Coalition, the South Asia Peasants Coalition, and the People's Coalition on Food Sovereignty, participated in the From Trade to Climate Caravan conducted prior to the Copenhagen COP15 mobilizations (Chatterton et al. 2013). Such solidarities are also expressed during the articulated moments of climate justice antagonism, such as the Climate Change, Gender and Food Sovereignty Caravan organized by the BKF, the BKS, and LVC in 2011. The Caravan was devised to educate and mobilize vulnerable peasant communities about the effects of climate change and facilitate movement-to-movement communication, the sharing of experiences, and the development of strategies. In so doing, it sought to deepen and extend networks of grassroots movements in South Asia concerning issues of climate change, gender, and food sovereignty. The Caravan comprised three buses containing 80 activists

(55 BKF/BKS activists from various districts of Bangladesh, and 25 activists from various international grassroots movements and groups[9]) engaging with BKF/BKS-organized peasant and indigenous communities through workshops, seminars, and rallies. The Caravan visited 18 villages in 12 districts of northern and southern Bangladesh. It had generative impacts on both intra-movement and inter-movement solidarities. According to one BKF activist, it helped to increase the organizational strength of the BKF and BKS across Bangladesh through the increased interaction between movement members from different parts of the country: "The Caravan was able to make a bridge between people in the north and south – who are facing different types of extreme weather events – to facilitate greater mutual understanding" (interview, Barisal District, Bangladesh, 2011). The participation of peasant activists from India, Nepal, Sri Lanka, Pakistan, and the Philippines also contributed to spatially extensive solidarity building with Bangladeshi peasants, as another BKF activist commented:

> The presence in communities of activists from other South Asian countries and from countries in the Global North was important in that it showed that the problems of those communities was of concern to others, and that the voices from the community were valued. This generated the feeling that local villagers were not alone in their struggle.
> *(Interview, Barguna District, Bangladesh, 2011)*

The Caravan also contributed to the fashioning of a counter-hegemonic common sense concerning agrarian issues and solidarity between movements. International participants interviewed on the Caravan felt that it provided an opportunity for activists to share experiences from their different movements' struggles and national contexts, explore how they might create longer-term solidarities (in particular bilateral campaigns with the BKF and BKS), fashion joint campaigns with other movements, and take their experiences back to their own countries and struggles. One Indian participant noted:

> The Caravan is a resource. We have formed relationships, deepened networking ties, especially with Pakistan, of which we had little knowledge before, and we have begun to plan future actions together. I think it was encouraging for communities to see an international presence, and that others care about the problems of people in Bangladesh and want to learn from them. This is solidarity. Bringing international folk onto the Caravan increases the image of the BKF on the ground. The Caravan provides the opportunity for the BKF to train local leaders in the key issues of the Caravan.
> *(Interview, Satkhira District, Bangladesh, 2011)*

Such "climate justice solidarities" (Routledge 2011) refer to how shared "maps of grievance" (Featherstone 2003) are constructed and link different activists involved in struggles over climate change. They bring together geographically, culturally, economically, and politically different and distant peoples and enable connections and alliances to be forged that extend beyond the local and particular (Olesen 2005). Events such as the Caravan generate connections between movements that vitiate against parochialism and chauvinism. Climate justice can function as a key discourse through which articulations are made between these diverse struggles.

In addition to reconfiguring some of the place-specific conditions of local organizing (strengthening the movement to help to ameliorate the impacts of government actors and the paucity of resources), the Caravan provided a productive space for generating future networking strategies; for example, planning meetings for launching a more extensive Caravan through India, Nepal, and Bangladesh in 2014 were held in Nepal in 2012 during a regional LVC gathering and in Jakarta, Indonesia, in 2013 during LVC's international conference.

Climate justice futures

This chapter has discussed discourses of climate justice emerging from civil society – in particular from social movements such as the BKF and BKS in the Global South. In examining the struggles for land waged by farmers' movements as they confront the challenges of climate change, I have argued that they amount to practices of climate justice. Indeed, in many ways land occupation is an attempt to materialize climate justice "on the ground," as, for poor farmers, it is a means to adapt to the challenges posed by climate change (interview, Dhaka, Bangladesh, 2011). Beyond Bangladesh, climate justice constitutes an important challenge to contemporary capitalist relations that involves three key dimensions. First, climate justice involves an antagonistic framing of climate politics that breaks with attempts to construct climate change as a "post-political" or "consensual" issue (Swyngedouw 2007, 2010). Rather than attempt to "reboot" capitalism, climate justice represents an unremitting critique of, and struggle against, business-as-usual and "green" forms of capitalist exploitation. By making such exploitation localizable and contestable, climate justice also poses alternatives to it. Hence, second, climate justice involves the formation of prefigurative political activity, especially in relation to commoning. This poses and practises collective alternatives to capitalist activities, for example in the form of food sovereignty, and in so doing brings other, more equitable and sustainable, worlds into being. Third, climate justice politics is generative of solidarities between differently located struggles both within countries and across national borders. The generation of such spatially extensive common ground is crucial, given that place-based communities and social movements are faced with powerful (corporate and government) opponents. As was seen with the alter-globalization struggles, such solidarities can pose significant challenges to further capitalist expansion.

Michael Watts (2013: xxx–xxxii) has recently argued that a marketized, global "new ecology of rule," driven by financial and development institutions, places the burden of climate change adaptation and survival even more on the backs of the poor. Certainly, activists within the BKF and BKS believe that they cannot rely on governments (or their corporate partners) to help them to overcome the economic and environmental challenges that they face. As a result, the politics of occupation remains a critical strategic resource and social justice response for peasant farmers in Bangladesh. Further, as the impacts of climate change intensify, the reframing of climate change as an issue of climate justice provides crucial alternative discourses, practices, and means of struggle against further rounds of capital accumulation.

Notes

1 www.internetpirate.com/Greenhouse%20Gangsters%20vs_%20Climate%20Justice.htm.
2 A network of over 160 organizations and networks (see www.climate-justice-now.org).
3 See www.ejnet.org/ej/bali.pdf; www.climate-justice-now.org; www.durbanclimatejustice.org; www.climate-justice-action.org; and http://unfccc.int/resource/docs/2010/awglca10/eng/misc02.pdf.
4 These include: the right to life and to exist; the right to water as a source of life; the right to clean air; the right to health; the right to be free from contamination, pollution, and toxic or radioactive waste; and the right not to have its genetic structure modified or disrupted.
5 Cyclone Sidr, in 2007, caused 3500 deaths (Karim and Mimura 2008).
6 See www.climatecamp.org.uk.
7 See www.coalactionscotland.org.uk.
8 For example, in Bangladesh, the BKF and BKS are members of the Aaht Sangathan, which also includes: the Floating Labour Union; the Floating Women's Labour Union; the Bangladesh Adivasi Samiti (Indigenous Committee); the Rural Intellectual Front; the *Ganasaya* Cultural Centre; and the Revolutionary Youth Association (interviews, Dhaka, Bangladesh, 2004, 2009).
9 Participation was from India (Andhra Pradesh Vyavasaya Vruthidarula Union; Karnataka State Farmers' Association; Institute for Motivating Self-Employment), Nepal (All Nepal Peasants' Federation; All Nepal Peasants' Federation (Revolutionary); All Nepal Women's Association; General Federation of Nepalese Trade Unions; Jagaran Nepal), Pakistan (Anjuman Muzareen Punjab (Tenants Association Punjab)), Sri Lanka (Movement for National Land and Agricultural Reform; National Socialist Party), and the Philippines (Kilusang Magbubukid ng Pilipinas (Peasant Movement of the Philippines; KMP)), as well as activists from La Via Campesina (South Asia), the UK, Germany, and Australia.

References

Alam, A., Shamsuddoha, M.D., Tanner, T., Sultana, M., Huq, M.J., and Kabir, S.S. (2011) "The political economy of climate resilient development planning in Bangladesh," *IDS Bulletin*, 42(3): 52–61.
Alam, B. (2008) "Food sovereignty and climate change," *The Struggle*, October: 4.
Alam, E. and Collins, A.E. (2010) "Cyclone disaster vulnerability and response experiences in coastal Bangladesh," *Disasters*, 34(4): 931–54.

Altieri, M.A. (2010) "Scaling-up agro-ecological approaches for food sovereignty in Latin America," in H. Wittman, A. Desmarais, and N. Wiebe (eds) *Food Sovereignty: Reconnecting Food Nature and Community*, Oxford: Pambazuka Press, pp. 120–33.

Ayers, J. and Huq, S. (2009) "Leading the way," *Himal Southasian*, October–November, http://old.himalmag.com/component/content/article/657-leading-the-way.html, accessed 17 December 2014.

Barthel, S., Crumley, C., and Svedin, U. (2013) "Bio-cultural refugia: Safeguarding diversity of practices for food security and biodiversity," *Global Environmental Change*, 23(5): 1142–52.

Brouwer, R., Akter, S., Brander, L., and Haque, E. (2007) "Socioeconomic vulnerability and adaptation to environmental risk: A case study of climate change and flooding in Bangladesh," *Risk Analysis*, 27(2): 313–26.

Building Bridges Collective (2010) *Space for Movement: Reflections from Bolivia on Climate Justice, Social Movements, and the State*, Leeds: Footprint Workers Co-op.

Carmody, P. (2009) "Cruciform sovereignty, matrix governance and the scramble for Africa's oil: Insights from Chad and Sudan," *Political Geography*, 28(6): 353–61.

Carroll, W.K. (2010) "Crisis, movements, counter-hegemony: In search of the new," *Interface: A Journal for and about Social Movements*, 2(2): 168–98.

Chatterton, P., Featherstone, D., and Routledge, P. (2013) "Articulating climate justice in Copenhagen: Antagonism, the commons, and solidarity," *Antipode*, 45(3): 602–20.

Chowdhury, A. (2009) "The coming crisis," *Himal Southasian*, October–November, http://old.himalmag.com/component/content/article/651-the-coming-crisis.html, accessed 17 December 2014.

Dasgupta, S., Huq, M., Khan, Z.H., Masud, M.S., Ahmed, M.M.Z., Mukherjee, N., and Pandey, K. (2011) "Climate proofing infrastructure in Bangladesh: The incremental cost of limiting future flood damage," *Journal of Environment and Development*, 20(2): 167–90.

della Porta, D., Andretta, M., Mosca, L., and Reiter, H. (2006) *Globalization from Below*, London: University of Minnesota Press.

Desmarais, A. (2007) *La Via Campesina: Globalization and the Power of Peasants*, London: Pluto Press.

Doyle, T. and Chaturvedi, S. (2011) "Climate refugees and security: Conceptualizations, categories, and contestations," in J. Dryzek, R. Norgaard and D. Schlosberg (eds) *The Oxford Handbook of Climate Change and Society*, Oxford: Oxford University Press, pp. 278–91.

Featherstone, D. (2003) "Spatialities of transnational resistance to globalization: The maps of grievance of the Inter-Continental Caravan," *Transactions of the Institute of British Geographers*, 28(4): 404–21.

Featherstone, D. (2005) "Towards the relational construction of militant particularisms: Or why the geographies of past struggles matter for resistance to neoliberal globalisation," *Antipode*, 37(2): 250–71.

Featherstone, D. (2008) *Resistance, Space and Political Identities: The Making of Counter-global Networks*, Oxford: Wiley-Blackwell.

Featherstone, D. (2012) *Solidarity, Hidden Histories, and Geographies of Internationalism*, London: Zed Books.

Federici, S. (2010) "Feminism and the politics of the commons in an era of primitive accumulation," in Team Colors Collective (ed.) *Uses of a Whirlwind: Movement, Movements, and Contemporary Radical Currents in the United States*, Edinburgh: AK Press, pp. 283–94.

Federici, S. (2012) *Revolution at Point Zero: Housework, Reproduction, and Feminist Struggle*, Oakland, CA: PM Press.

Franks, B. (2003) "Direct action ethic," *Anarchist Studies*, 11(1): 13–41.
Giddens, A. (2009) *The Politics of Climate Change*, Cambridge: Polity.
Gilman, N., Randall, D., and Schwartz, P. (2011) "Climate change and security," in J. Dryzek, R. Norgaard, and D. Schlosberg (eds) *The Oxford Handbook of Climate Change and Society*, Oxford: Oxford University Press, pp. 251–66.
Gilroy, P. (2010) *Darker than Blue: On the Moral Economies of Black Atlantic Culture*, Cambridge, MA: Belknap Press.
Gramsci, A. (1971) *Selections from Prison Notebooks*, New York: International Publishers.
Hardt, M. and Negri, A. (2009) *Commonwealth*, London: Harvard University Press.
Harvey, D. (2003) *The New Imperialism*, Oxford: Oxford University Press.
Holt-Gimenez, E. and Patel, R. (2009) *Food Rebellions! Crisis and the Hunger for Justice*, Oxford: Pambazuka Press.
Hossain, M. (2009) "Dynamics of poverty in rural Bangladesh 1988–2007: An analysis of household level panel data," paper presented at the Conference on Employment Growth and Poverty Reduction in Developing Countries, Political Economy Research Institute, University of Massachusetts, Amherst, USA, 27–28 March, www.peri.umass.edu/fileadmin/pdf/conference_papers/khan/Hossain_Bangladesh.doc&sa=U&ei=qKWRVOmML8yz7gbS3IDgCw&ved=0CAYQFjAA&client=internal-uds-cse&usg=AFQjCNGacNEPxtXDishjViNKou3fSAr _bg, accessed 17 December 2014.
Intergovernmental Panel on Climate Change (IPCC) (2008) *Climate Change 2007 Synthesis Report*, Sweden: Teri Press.
Juris, J. (2007) "Practicing militant ethnography with the movement for global resistance," in S. Shukaitis and D. Graeber (eds) *Constituent Imagination: Militant Investigation, Collective Theorization*, Oakland, CA: AK Press, pp. 164–76.
Juris, J. (2008) *Networking Futures*, Durham, NC: Duke University Press.
Karim, M.F. and Mimura, N. (2008) "Impacts of climate change and sea level rise on cyclonic storm surge floods in Bangladesh," *Global Environmental Change*, 18: 490–500.
Karriem, A. (2009) "The rise and transformation of the Brazilian landless movement into a counter-hegemonic political actor: A Gramscian analysis," *Geoforum*, 40: 316–25.
Katz, C. (2001) "On the grounds of globalization: A topography for feminist political engagement," *Signs*, 26(4): 1213–29.
Klare, M. (2001) *Resource Wars*, New York: Metropolitan.
La Via Campesina (2009) "Small scale sustainable farmers are cooling down the earth," Jakarta: La Via Campesina, http://viacampesina.org/downloads/pdf/en/EN-paper5.pdf, accessed 17 December 2014.
Le Billon, P. (2004) "The geopolitical economy of 'resource wars,'" *Geopolitics*, 9(1): 1–28.
Levien, M. (2013) "The politics of dispossesion: Theorizing India's 'land wars,'" *Politics and Society*, 41(3): 351–94.
Martinez-Alier, J. (2002) *The Environmentalism of the Poor: A Study of Ecological Conflicts and Valuation*, Cheltenham: Edward Elgar.
Martinez-Alier, J. and Temper, L. (2007) "Oil and climate change: Voices from the South," *Economic and Political Weekly*, 42(50): 16–19.
Martinez-Alier, J., Anguelovski, I., Bond, P., Del, Bene, D., Demaria, F., Gerber, J.-F., Greyl, L., Haas, W., Healy, H., Marin-Burgos, V., Ojo, G., Firpo Porto, M., Rijnhout, L., Rodriguez-Labajos, B., Spangenberg, J., Temper, L., Warlenius, R., and Yánez, I. (2014) "Between activism and science: Grassroots concepts for sustainability coined by environmental justice organizations," *Journal of Political Ecology*, 21: 19–60.
Mountz, A. (2013) "Political geography I: Reconfiguring geographies of sovereignty," *Political Geography*, 37(6): 829–41.

Muradian, R. and Martinez-Alier, J. (2001) "Trade and the environment: From a 'southern' perspective," *Ecological Economics*, 36(2): 281–97.

Murshid, K.A.S. (n. d.) *Implications of Agricultural Policy Reforms on Rural Food Security and Poverty*, www.saprin.org/bangladesh/research/ban_agri_policy.pdf, accessed 17 December 2014.

Nally, D. (2011) "The biopolitics of food provisioning," *Transactions of the Institute of British Geographers*, 36: 37–53.

Olesen, T. (2005) *International Zapatismo*, London: Zed Books.

Panitch, L. and Gindin, S. (2011) "Capitalist crises and the crisis this time," *Socialist Register*, 47: 1–21.

Parenti, C. (2011) *Tropic of Chaos: Climate Change and the New Geography of Violence*, New York: Nation Books.

Patel, R. (2009) "What does food sovereignty look like?," *Journal of Peasant Studies*, 36(3): 663–73.

Paul, B.K. and Dutt, S. (2010) "Hazard warnings and responses to evacuation orders: The case of Bangladesh's Cyclone Sidr," *Geographical Review*, 100(3): 336–55.

Raihan, M.S., Huq, M.J., Alsted, N.G., and Andreasen, M.H. (2010) *Understanding Climate Change from Below, Addressing Barriers from Above: Practical Experience and Learning from a Community-based Adaptation Project in Bangladesh*, Dhaka: ActionAid.

Rosset, P.M. (2003) "Food sovereignty: Global rallying cry for farmer movements," *Food First Backgrounder*, 9(4): 1–4.

Rosset, P.M., Sosa, B.M., Jaime, A.M.R., and Lozano, D.R.A. (2011) "The *campesino*-to-*campesino* agroecology movement of ANAP in Cuba: Social process methodology in the construction of sustainable peasant agriculture and food sovereignty," *Journal of Peasant Studies*, 38(1): 161–91.

Routledge, P. (1996) "The third space as critical engagement," *Antipode*, 28(4): 397–419.

Routledge, P. (2002) "Travelling east as Walter Kurtz: Identity, performance and collaboration in Goa, India," *Environment and Planning D: Society and Space*, 20: 477–98.

Routledge, P. (2003) "Convergence space: Process geographies of grassroots globalization networks," *Transactions of the Institute of British Geographers*, 28: 333–49.

Routledge, P. (2008) "Acting in the network: ANT and the politics of generating associations," *Environment and Planning D: Society and Space*, 26(2): 199–217.

Routledge, P. (2011) "Translocal climate justice solidarities," in J. Dryzek, R. Norgaard, and D. Schlosberg (eds) *The Oxford Handbook of Climate Change and Society*, Oxford: Oxford University Press, pp. 384–98.

Routledge, P. and Cumbers, A. (2009) *Global Justice Networks: Geographies of Transnational Solidarity*, Manchester: Manchester University Press.

Seabrook, J. (2013) "In the city of hunger: Barisal, Bangladesh," *Race and Class*, 51(4): 39–58.

Swyngedouw, E. (2007) "Impossible 'sustainability' and the post-political condition," in D. Gibbs and R. Krueger (eds) *The Sustainable Development Paradox: Urban Political Economy in the United States and Europe*, New York: Guilford Press, pp. 13–40.

Swyngedouw, E. (2010) "Apocalypse forever? Post-political populism and the spectre of climate change," *Theory, Culture & Society*, 27(2–3): 213–32.

Watts, M. (2013) *Silent Violence: Food, Famine, and Peasantry in Northern Nigeria*, London: University of Georgia Press.

Wilkinson, R. and Pickett, K. (2009) *The Spirit Level: Why More Equal Societies Almost Always Do Better*, London: Allen Lane.

Windfuhr, M. and Jonsen, J. (2005) *Food Sovereignty: Towards Democracy in Localised Food Systems*, Warwickshire: ITDG.

Wittman, H. (2009) "Reworking the metabolic rift: *La Via Campesina*, agrarian citizenship, and food sovereignty," *Journal of Peasant Studies*, 36(4): 805–26.

Wittman, H. (2010) "Reconnecting agriculture and the environment: Food sovereignty and the agrarian basis of ecological citizenship," in H. Wittman, A. Desmarais, and N. Wiebe (eds) *Food Sovereignty: Reconnecting Food, Nature, and Community*, Oxford: Pambazuka Press, pp. 91–105.

Wolford, W. (2010) *This Land Is Ours Now: Social Mobilization and the Meanings of Land in Brazil*, Durham, NC: Duke University Press.

Wright, E.O. (2010) *Envisioning Real Utopias*, London: Verso.

6
CLIMATE CHANGE AND THE INSECURITY FRAME

Simon Dalby

Environmental security and climate conflict

Environmental change has long been discussed as a threat to many things. In the late 1980s, as the Cold War wound down, alarms about environmental matters were high-profile issues. Not surprisingly, policy-makers and academics began to ponder new security agendas once the major threat of nuclear war had faded (Mathews 1989). The hot summer of 1988 in the United States, widespread deforestation in the Amazon, and worries about stratospheric ozone depletion and the aftermath of the Chernobyl nuclear reactor meltdown, as well as the Bhopal chemical poisoning disaster in India, focused attention on environmental matters. Given that security, or – in the case of the United States, in particular – national security, is the highest priority of state institutions, it is not surprising that the language of security was invoked to give environmental matters priority on the post-Cold War policy agenda and since (Floyd and Matthew 2013).

Now, a quarter of a century later, this environmental security debate has been re-energized with the focus much more explicitly on climate change. A whole "climate security" discourse has emerged, mostly since 2007, that links energy, environmental change, climate, and appeals to national security, in the United States in particular, and to various other formulations of security elsewhere (Webersik 2010). The crucial point is that climate change is about much more than traditional environmental concerns with parks, pollution, preservation, and population. According to those who specify climate change in terms of national or human security, it is a much more important issue, potentially leading to major social disruptions and possibly wars, and hence a much higher policy priority.

This high priority involves using the rhetoric of threats to "securitize" the issue in terms of making it a priority that rises above "normal" politics or routine administration. Securitization analyses suggest that matters become "security" issues when political discourse invokes dangers to some referent object; threats apparently require extraordinary measures beyond the normal, routine operations of politics within states (Buzan et al. 1998). There is no guarantee that all claims of imminent danger or even existential threats to a state will lead to a successful securitization of an issue, nor to the effective formulation or execution of a security strategy to deal with the threat (Fierke 2007). Likewise, securitizations take place in complex political landscapes where multiple dangers shape policy and political actions, and where prior security priorities or immediate political interests may trump new ones, even those with apparently drastic consequences.

As critics of environmental security have been pointing out for a long time, however, simply raising the issue to a matter of high priority might not be helpful if the policy instruments used to address the matter are inappropriate (Deudney 1990; Finger 1991). At least in the case of the military dimensions of security thinking, it is far from clear that the military is the appropriate institution to address climate change. When at war or training for war, the armed forces are usually better known for destroying environments. That said, in terms of early warning, monitoring, and preparing to deal with environmental disruptions, the military is often the lead agency coping with such things and a useful partner for numerous civilian research initiatives (O'Lear et al. 2013). These contradictions emphasize the importance of being clear about what is a threat to whom, where, and how such threats should be addressed. If security is linked to some notion of global or human security, rather than the national security of the more powerful industrial states, the policy discourse might be much more appropriate to climate issues generally (Dalby 2009). How climate change is framed in relation to various notions of security is key to this discussion.

To tease out the importance of framing climate in relation to various formulations of security, this chapter focuses first on a key report issued in 2007 by the CNA Corporation (a defence think-tank that had its origins in the earlier Center for Naval Analysis in Washington), where climate was first publicly framed as a security threat because of its potential as a "threat multiplier" that might aggravate political instabilities. The "threat multiplier" formulation in this report was subsequently used two years later in the United Nations Secretary General's report on climate and security. An update of the original 2007 report drafted by the CNA Military Advisory Board in 2014 suggested that American national security needs to be substantially rethought in the face of climate change. Crucially, however, the 2014 version of the report forcefully framed climate as a matter for the present, a catalyst for conflict, not – as had been the case in the 2007 report – as a threat multiplier that mostly applies to future conflicts. That future has now arrived, the most recent report implied.

Nevertheless, just as such blunt assessments of imminent danger were being widely publicized in Washington, they were being simultaneously rejected as political ideology by politicians – mainly Republicans, but a few Democrats too – who represented districts with large oil and coal production. These politicians were apparently anxious to deflect any policy initiatives or federal spending that might impede the rush to frack oil and gas fields and mine coal in the United States. Where enthusiastic support for security and military matters might once upon a time have been taken for granted in the United States, the politics of climate change has pitted military leaders against some of their traditional political supporters in Congress. In 2014, the attempt to securitize climate change in the latest CNA report, and many related opinion pieces in the media published by senior members of the national security establishment, was flatly rejected by several politicians who might reasonably have been expected to be sympathetic to such a framing.

National security and the threat of climate change

As Andrew Szasz outlines in Chapter 10 of this volume, the US military has taken a growing interest in climate change over the last decade. While a flurry of media attention was paid to a 2003 scenario document prepared by some consultants that purported to show that climate was a likely key source of future conflict (Schwartz and Randall 2003), this was just one more scenario exercise rather than a major institutional initiative. In the aftermath of 9/11, the Bush administration was seriously opposed to linking climate concerns with national security, and reluctant to commission any official US military publications on the theme. The report that finally broke the silence on climate and security and focused attention on the issue was the aforementioned CNA Corporation's *National Security and the Threat of Climate Change* (2007). This was quickly followed by similar reports from the Center for Strategic and International Studies (Campbell et al. 2007) and the Strategic Studies Institute of the American War College (Pumphrey 2008). A German Advisory Council on Global Change report on the security risks of climate change also appeared in 2007, with an English translation pubilshed the following year (WBGU 2008). While 2007 was also the year in which the Intergovernmental Panel on Climate Change (IPCC) presented its *Fourth Assessment Report* and shared the Nobel Peace Prize with Al Gore (for his *An Inconvenient Truth* documentary), it is important to note that the IPCC did not discuss climate in terms of security in its assessment.

Seven years later, as the Obama administration made climate change a key part of the plans for its legacy, and military planning to deal with both increasingly severe disasters and the vulnerability of its own facilities moved ahead, the CNA Military Advisory Board (2014) issued an updated report on the national security implications of climate change. It urged immediate action on climate change; without this, the authors warned, various calamities might result. As with the first report, the 2014 version once again quoted senior retired military figures discussing the practicalities of military operations and the need for carefully thought-through risk assessments in planning to add weight to the analysis. However, this report, unlike the first, was also able to draw on a series of national climate assessments (Melillo et al. 2014) and other official documents to add to the arguments about why climate is increasingly a matter of prime importance to many dimensions of American policy. The response from many Republican politicians in Washington to such arguments in 2014 was in part an attempt to direct the military to ignore climate change in its planning. It is not, so the counter-argument went, directly about military supremacy or war fighting; hence, it is a political distraction to the core military interests of national security.

Is climate change, then, to be understood as a security issue or not? How it is framed is both a conceptual matter and a practical matter of partisan politics in Washington in particular. It is also clearly a matter of how security is understood, and how and where dangers are specified in both policy and academic texts that link climate change to all manner of political matters. In 2014, too, the IPCC finally included a discussion of security in its *Fifth Assessment Report*, but one more closely linked to matters of human rather than national security. Vulnerabilities to storms and climate disruptions, and the need to adapt to changing circumstances, are key to these formulations, rather than military preparations or "national security." As with the larger discussion of human security, the IPCC summation is more about preparation and anticipating likely dangers than it is about heroic responses once disaster strikes.

If climate is to be linked to security, the question of how these links are to be made is therefore very important, both as an analytical matter and as a matter of policy formulation. Crucial to the CNA Corporation report in 2007, although not developed in any analytical depth, is the formulation of climate change as a "threat multiplier." The opening address to readers is quite specific: "Climate change can act as a threat multiplier for instability in some of the most volatile regions of the world, and it presents significant national security challenges for the United States" (CNA Corporation 2007: 1). The executive summary reprints these words and then elaborates on the logic:

> In the national and international security environment, climate change threatens to add new hostile and stressing factors. On the simplest level, it has the potential to create sustained natural and humanitarian disasters on a scale far beyond those we see today. The consequences will likely foster political instability where societal demands exceed the capacity of governments to cope.
>
> *(CNA Corporation 2007: 6)*

Hence, the argument goes, climate change adds to existing security problems and as such needs to be taken seriously by security planners both in anticipating likely future trouble spots and in preparing to deal with humanitarian disasters and political difficulties. This logic leads to the conclusion that the US military needs to incorporate climate change into its planning and think through the vulnerability of its facilities and operations to extreme weather events.

Not worked out in any detail in the report are the precise ways in which climate might cause greater instability, nor how, having done so, this might aggravate political difficulties in a way that makes situations amenable to exploitation by extremist forces, terrorists, or insurgencies. The assumption that these matters will be aggravated by weather disruptions is "common sense" and taken for granted in a view of the world that suggests that stability is key to security; change that reduces the efficacy of conventional modes of governance is inevitably a problem for the larger social order. While disasters are obviously an immediate issue for governments, what is not clear in this formulation is that their response to a crisis might be far more important than the crisis itself in terms of whether threats multiply. That conflict might come from the mishandling of a short-term weather-related disaster, such as a hurricane, or a longer-term drought event that disrupts water supplies and agriculture, is not discussed in much detail. As subsequent events in Syria in particular have made clear, this is a crucial dimension to the linkage of climate to conflict (Gleick 2014). Likewise, the prior existence of structural injustices in the access to resources – a matter that frequently underlies the outbreak of violence when weather and/or agricultural disruptions heighten problems of sustenance – is not countenanced in these formulations of security difficulties (Zografos et al. 2014).

The CNA Corporation report in 2007 also made interesting observations about the benefits of military uses of renewable energy sources in preference to the traditional reliance on fossil fuels, noting that the US forces' supply lines in both Iraq and Afghanistan were especially vulnerable because of the vast amounts of fuel that had to be transported to forward operating locations, and the susceptibility of these convoys to attack. Hence, the CNA Corporation report argues that solar power might be a practical asset to the military as well as an effective strategy to reduce the speed of climate change. Such reasoning works well in military circles where operational flexibility is important and logistical vulnerabilities are always a concern to planners. But it is not an argument likely to win much sympathy from traditional environmentalists anxious to emphasize the damage done by military action and by industrial-scale preparations and training exercises for combat operations.

The United Nations and climate security

Concerns about climate change finally made their way onto the agenda of the United Nations Security Council in 2009, driven by European alarm at the slow pace of international progress on climate negotiations. The background report by the Secretary General (UN 2009) to the General Assembly draws on the threat multiplier formulation in its summary of the links between security and climate:

> Climate change is often viewed as a "threat multiplier," exacerbating threats caused by persistent poverty, weak institutions for resource management and conflict resolution, fault lines and a history of mistrust between communities and nations, and inadequate access to information or resources.
> *(UN 2009: 2)*

The response suggested in policy terms involves a series of threat minimizers, formulated to help states deal with climate difficulties and security issues related to climate. The lack of specificity is perhaps unsurprising, given that this is a UN document. The threat minimizers include:

> climate mitigation and adaptation, economic development, democratic governance and strong local and national institutions, international co-operation, preventive diplomacy and mediation, timely availability of information and increased support for research and analysis to improve the understanding of linkages between climate change and security. Accelerated action at all levels is needed to bolster these threat minimizers.
> (UN 2009: 2)

The UN report also provides a list of possible developments that will require the attention of the international community as climate change continues to unfold, including "loss of territory, statelessness and increased numbers of displaced persons; stress on shared international water resources, for example, with the melting of glaciers; and disputes surrounding the opening of the Arctic region to resource exploitation and trade" (UN 2009: 2). Clearly, all of this is premised on the international community holding serious international negotiations to reduce greenhouse gas emissions, without which any other policy measures would be, at best, palliative actions.

The UN report draws on submissions from member states that, in light of the Millennium Development Goals and earlier formulations of human security, mainly emphasize the importance of security threats to individuals. A section from Chapter 2 of the *Human Development Report* (UN 1994) – the document that explicitly formulated human security for the UN community – is reproduced verbatim: "security symbolized protection from the threat of disease, hunger, unemployment, crime, social conflict, political repression and environmental hazards" (UN 2009: 5). While it is recognized that these threats might have effects on national security in particular states, this theme is not given the emphasis that the earlier US documents tied into national security and military preparation suggest. The emphasis on the vulnerability of peoples rather than the disruption of political order is noteworthy; the link is explicitly made to matters of development, the Millennium Development Goals, and related diplomatic initiatives, not to the national security of a particular state.

Five "channels" are identified in the Secretary General's report that link climate change as a threat multiplier to social phenomena. These draw on the diverse submissions from national delegations on the theme of climate change. First, climate change impacts on the well-being of the most vulnerable communities. Second, the report notes that climate change may disrupt economic growth and, in the process, increase desperation among impoverished peoples, with political disruptions potentially following. Third is the matter of uncoordinated and failed attempts to adapt to climate change that might lead to conflict or even resource wars, not least because of human migration. Fourth, some submissions to the report explicitly raise the question of the physical survival of some states, in particular small, low-lying island states that face eradication by rising sea levels. Finally, the potential for resource disputes and territorial conflict as a result of changes in natural resource vulnerabilities, and such issues as access to resources in the Arctic, is noted with concern.

The bulk of the report itemizes these channels in some detail, emphasizing that there is a widespread consensus that growth is seen as the key to dealing with any difficulties that might be caused (even if only indirectly) by climate change. The report draws heavily on IPCC documents and the *Fourth Assessment Report* (IPCC 2007) in particular. It notes the difficulty of teasing out causal and contextual issues in distinguishing what climate change might cause as distinct from difficulties faced by societies regardless of global warming. The upshot is that this report suggests that sustainable development and poverty alleviation are key to giving states and peoples the tools they need to cope with a changing climate. Quite why threat multipliers and channels are the most appropriate framework for this UN document is not always clear, but their adoption seems to be directly related to attempts to frame matters of climate in terms of security. Because this is a UN report on the security dimensions of climate change, this framing was the obvious one that was available in US policy circles in 2009 to link climate to traditional notions of security concerning violent conflict. It was published in the aftermath of a flurry of publications in 2007 and 2008 that frequently used threat multipliers and similar formulations. However, in contrast to the UN report, most of these earlier documents were linked to US national security, not to larger concerns with vulnerable populations or human security.

Clearly, there are some circumstances where climate change could cause domestic political instabilities, with the potential to develop into international conflicts, but mostly the Secretary General's report suggests that climate change is a matter of making development more difficult and primarily, despite the language of security, a matter of sustainable development more directly. That said, the original formulation of human security as the provision of the conditions for development underlies much of this logic (UN 1994), even without the explicit formulation in terms of threat multipliers. In short, if climate is to be reframed as a matter of security rather than as an environmental matter, then it seems that the UN is prioritizing formulations in terms of human security rather than traditional matters of national security among its members. There is a very obvious tension between the formulation that will resonate with Washington policy-makers focused on US concerns with political instability in various places, and the wider international community worried that climate change will interfere with development objectives in poor countries.

There is little in this report to suggest explicitly that climate will cause warfare, but much to suggest that instabilities are likely to be exacerbated by climate change, making sustainable development, the *sine qua non* for other matters, more difficult to achieve. In this, the report fits well with the subsequent scholarly analysis of current climate difficulties and the importance of linking environmental change to larger discussions of conflict and human security (O'Brien et al. 2013). It is not about national security; its implicit geography is about multiple difficulties in many places, not about specific threats to particular states that require national security and unilateral action by any single state. Framing matters in terms of threats to national security is at best a poor fit outside the Washington policy community. And, as the next two sections of this chapter illustrate, the national security framing did not work well even in Washington in subsequent years.

Updating the CNA analysis: accelerating threats

Five years after the Secretary General's report and shortly after the publication of the IPCC's *Fifth Assessment Report* and the third US national climate assessment (Melillo et al. 2014), the CNA Military Advisory Board (2014) published an update of the corporation's earlier report. Once again, the report emphasizes the practical experience of retired senior military officers in making its assessment of what are now termed the "accelerating risks" of climate change. The foreword is signed by Michael Chertoff, former Secretary of Homeland Security, and Leon Panetta, former US Secretary of Defense, to emphasize the gravity of the implications of the analysis that follows. Key to the 2014 report is the assertion that climate change is already happening and having effects. It is not a matter for the future. It is happening today and requires a bipartisan policy response in Washington. The report explicitly states its authors' dismay at the delay in addressing climate change mitigation internationally, not least because matters are progressing more quickly than their earlier report suggested might be the case. Now climate is framed as an urgent and immediate problem, not a matter for the future.

While the formulation of threat multipliers is again invoked in the 2014 report, the key findings suggest that something more direct than multipliers is already happening. Now, while climate ought to be a matter for international cooperation, it appears to be a catalyst for conflict:

> In Africa, Asia, and the Middle East, we are already seeing how the impacts of extreme weather, such as prolonged drought and flooding – and resulting food shortages, desertification, population dislocation and mass migration, and sea level rise – are posing security challenges to these regions' governments. We see these trends growing and accelerating.
> (CNA Military Advisory Board 2014: 2)

Moreover, it is clear that the report's authors are worried that people are failing to consider what is coming. Growing coastal settlements and population movements to cities, combined with the global interconnections that are making environmental policy ever more complicated, suggest to the authors that the world must guard against a failure of imagination. While this is not new in military terms – unanticipated dangers are key to military history, after all – the authors are clearly concerned that policy-makers and citizens are simply not thinking seriously about the processes already in motion and the need to adapt to them while simultaneously slowing climate change and making societies more resilient to climate shocks. Pointedly, the report notes that the failure to anticipate the 9/11 attacks was in part due to a lack of appropriate imagination and insufficient attention to trends around the world. The imputation is that policy elites are once again failing to think carefully about how the world is being changed. If this continues, nasty surprises will result.

More so than in the 2007 document, the authors of the 2014 report are concerned that climate change has the potential to degrade national power and the flexibility the United States needs to deal with international matters. In Pentagon acronym terms, national power is the ability to exert influence by diplomatic, information, military, and economic (DIME) means. Climate change may reduce this ability by degrading what are called political, military, economic, social, infrastructure, and information (PMESII) systems. While this jargon may be peculiar to military thinking, the focus on reduced capabilities is clear. All of this matters now because of the global interconnections between various parts of the world, a complicated situation that the authors suggest demands more attention than the earlier report recommended.

This shift in geopolitical sensibility, with its emphasis on global interconnections, is a noteworthy development in the latter report. Geopolitical stability is specified as the key to national security, and this is now in danger in ways that analysts are only beginning to contemplate. The more complicated interconnections between climate and stability are discussed with reference to attempts to think through some of the factors that influenced, if not directly caused, the political changes in the Middle East during the "Arab Spring." Rising global food prices, caused in part by drought conditions in China and Russia, aggravated social tensions in the Arab world, which in turn fed the turmoil. Hence, the report warns that global interconnections matter in terms of stability and potential conflicts. Syria's civil war was preceded by years of drought that triggered migration in the country that the government did not deal with effectively; in turn, this resulted in the galvanization of political opposition. Clearly, in this case, climate may be understood as a catalyst of warfare.

Disruptions to supply chains in the global economy are also noted as problematic for many states and corporations. Cargo ships are beginning to use the Arctic as a shipping lane as the ice continues to recede dramatically; the first commercial voyage occurred in September 2014, just a few months after the CNA Military Advisory Board report was published. Such developments are the new geopolitical context in which security understood as stability now has to be reconsidered. While the term "globalization" appears nowhere in the report (and the adjective "globalized" only twice), the military minds that drafted it clearly believe that this is the context in which instabilities now have to be considered.

As the report was prepared by retired military experts, a lengthy section on forces training, readiness, and capabilities was inevitable. So too was the declaration that climate is not the only security threat that the military might have to face in the foreseeable future. Nevertheless, the report notes that climate change may have unforeseen impacts on everything from training schedules to deployment. If droughts exacerbate fire dangers on training grounds, live-fire exercises may have to be curtailed. If National Guard units are deployed to aid disaster-stricken areas in the United States, they will be unavailable for use elsewhere. The vulnerability of bases to storm surges in coastal areas may become a factor in deciding which facilities are closed or remodelled in future infrastructure discussions. In short, climate will have multiple effects on military operations, in addition to being a major factor in the larger geopolitical situation. While there are other dimensions to national power and PMESII, clearly, the report's authors argue, climate change now has to be included in the planning of all manner of DIME policies. The Pentagon's subsequent plan for climate change adaptation follows this advice and incorporates most of these themes (US Department of Defense 2014).

US politics and climate change

While the CNA Military Advisory Board (2014) may be alarmed at the failure of national planning in the United States and the lack of substantial progress in international negotiations on climate change, the political opposition to dealing with climate change has persisted in Washington and elsewhere in the country to a degree that requires some explanation. The science is clear. The projections of what might be coming are fairly robust. The empirical connections between climate change and the consequences of lack of preparation to deal with it are also becoming increasingly apparent. Yet, political opposition to planning for a rapidly changing world persists. In the United States, the Republican Party has repeatedly expressed active opposition to climate-related planning and attacked what it terms Obama's "war on coal." These campaigns have long involved organized efforts to deny the legitimacy of climate science (Jacques et al. 2008; Dunlap 2014), and these have been extended by a proliferating set of internet sites and across numerous other media outlets (Sharman 2014). Various motions and pieces of legislation aimed at explicitly preventing public agencies from planning on the basis of climate change predictions have been passed in recent years. Clearly, there is concerted rejection of the climate security framing of public policy; apparently, climate change is not a matter of human actions and hence should not feature in discussions of security.

There are various explanations for this, related to the "culture wars" that are occurring in the "Anglosphere" of Australia, Canada, the United Kingdom, the United States, and perhaps New Zealand. Fossil fuel industries and reactionary politicians opposed to state initiatives to promote renewable energy or actively encourage innovations to reduce fossil fuel use have maintained a political campaign to obstruct innovation while simultaneously supporting continued subsidies for oil exploration and extraction. Hype about the supposedly huge potential for energy supplies made possible by fracking has also fed into related arguments that energy security is most important in a dangerous geopolitical environment where dependence on imported fuel supposedly makes the United States vulnerable (Yergin 2011). There are extensive links between the fossil fuel industry and US electoral politics that are beyond the scope of this chapter, but it is a very important factor especially in the recent growth of US hydrocarbon production and in the environmental struggles to try to prevent fracking and other destructive modes of extraction (Klein 2014).

One episode in particular highlights how the climate security frame has been rejected. In the aftermath of the publication of the 2014 CNA Military Advisory Board report, a House of Representatives motion to amend a budget bill to fund the Department of Defense was passed. This bill specifically stated that the Pentagon should not spend funds on climate matters:

> None of the funds authorized to be appropriated or otherwise made available by this Act may be used to implement the US Global Change Research Program National Climate Assessment, the Intergovernmental Panel on Climate Change's *Fifth Assessment Report*, the United Nations' Agenda 21 sustainable development plan, or the May 2013 *Technical Update on the Social Cost of Carbon for Regulatory Impact Analysis – Under Executive Order 12866*.
> (McKinley 2014)

In a letter urging support for his motion, the instigator of this bill, the Republican David McKinley, declared:

> Our climate is obviously changing; it has always been changing. With all the unrest around the global [*sic*], why should Congress divert funds from the mission of our military and national security to support a political ideology? ... [T]his amendment will ensure we maximize our military might without diverting funds for a politically motivated agenda.
> (Quoted in Gutman 2014)

Opponents of climate change actions introduced many other motions in attempts to prevent serious planning to deal with climate disruptions in 2014. They formed part of a campaign that continued through much of the year, leading up to the mid-term elections. However, McKinley's motion is particularly noteworthy because he explicitly framed climate as ideology rather than science, and hence as something to be dismissed. He did this by implying that both the UN and the IPCC are parts of a political agenda, not dealing with science. His reference to the IPCC's *Fifth Assessment Report* (2014) is less than clear – as this report does not advocate any specific implementation activities – but it certainly imputes a political rather than scientific origin to the IPCC's work. Linked to his reference to the *Technical Update on the Social Cost of Carbon* (Interagency Working Group on the Social Cost of Carbon 2013), this clearly ties the Obama administration's efforts to tackle climate change directly to the IPCC and the UN, hence suggesting that they all share a single agenda. Climate change is now reframed in these terms as political ideology, which apparently is not the basis for national security action.

The links between climate and insecurities are flatly denied in McKinley's formulation. Unrest around the globe is defined as something unrelated to climate. McKinley implies that political motivations are behind such unrest, rather than a rapidly changing world in which some people are made especially vulnerable by severe weather, droughts, and related disruptions to life's routines. Here, he flies in the face of shifting discourses of climate security in other developed countries, where the complex interactive dimensions of insecurity are increasingly understood as key to policy formulation (Mayer 2012). The implication is that the United States stands apart from many of these other problems. Wars and disruptions are either part of the human condition or take place beyond US shores; they are certainly not caused by US consumption of fossil fuels. The implicit geography of national security, of a separate entity facing external threats, starkly contrasts with the global premises of most of the UN discussion and human security generally (Scheffran et al. 2012).

Particularly interesting in McKinley's formulation, and noteworthy in terms of how climate is framed, is his insistence that climate "has always been changing." Whether humanity causes this change does not matter here; the "denial" of anthropogenic change – a standard argument against reducing carbon emissions – is implied, but not stated directly. Insofar as there are arguments in the climate security discussion that relate to military operations, regardless of the imputed causality, these are simply not to be funded. Environmental change is understood simply as the given backdrop of human affairs, not as something that might be caused by human actions or, even if it were, as something that should be considered as an appropriate matter for policy attention or practical action by the armed forces.

Security is not related to environmental matters in McKinley's formulation, nor to the larger, complicated matters of economic and political globalization that are considered in the 2014 CNA Military Advisory Board report. This contextualization is simply rejected. The implicit geographical assumption that underlies McKinley's motion is simply of a given *context* for security action; he feels that defence planners can do nothing to tackle – nor even adequately prepare for – this. His logic fits with contemporaneous attempts by Australian and Canadian federal politicians to stymie international efforts to tackle climate change (Dalby forthcoming). However, the blunt denial that climate may have any connection to insecurity is a noteworthy political attempt to refuse the securitization of climate change.

As the securitization discussion in the scholarly literature makes clear, attempts to render matters in the security frame are not always successful; this contestation emphasizes this point. The contrast between the dismissal of environmental and related globalization matters as of security concern and the alarm expressed especially by the CNA Military Advisory Board authors is stark. It emphasizes the crucial point that framing political and security matters is important, but the successful climatization of national security discourse is far from a foregone conclusion (Oels 2012). Crucial parts of the Republican audience in the United States have simply rejected it, and did so to good effect for their party in the mid-term elections of November 2014 when a number of high-profile critics of climate change action were elected.

Anthropocene geopolitics?

The contrast between the logic of McKinley's motion and the invocation of climate as a matter of security is even more dramatic when compared with the rapidly growing and robust earth system science analyses that have appeared in the last decade (Whitehead 2014). These focus on climate change as but one of the transformations of the earth system that humanity has set in motion. They require recognition of humans as ecological change agents on a global scale and, as such, effectively a new geological factor in planetary history. Hence the now frequent invocation of the term "Anthropocene" to suggest these new circumstances. This science requires a rethinking of modernity's premise of a separate sphere of human and natural phenomena and a clear understanding of complex relationships and causes in the system as a whole as the premise for thinking and action. Autonomous states, peoples, societies, or territories simply do not work as the appropriate premise for action in this new understanding of the human condition (Dalby 2014).

Such formulations are anathema to Republican politicians in Washington, for whom such science is apparently a political ideology, not a matter of reliable knowledge that can be accepted as the basis for action. There are other reasons for rejecting climate science, not least because it may work through environmental policy to constrain the actions of fossil fuel companies, but the point for this chapter is that the *framing* as science is simply rejected. This denial of the larger context of global interconnections, and with it policies to facilitate the rapid de-carbonization of the global economy (the *sine qua non* of dealing with long-term climate change), starkly illustrates the importance of framing in this case. It simultaneously emphasizes the contested nature of science in the formulation of public policy where it might impact the profitability of the fossil fuel industry. The Anthropocene formulation suggests powerfully that human activities have consequences for the future of the planet that demand urgent attention (Steffen et al. 2011). Dismissing such arguments – and assuming that the geographical context is simply a given rather than one that is being rapidly reshaped by human activity – will probably result in a more crisis-ridden, unstable, and potentially violent future than would otherwise be the case.

As the Secretary General's report (UN 2009) analysed here also suggests, climate security is a matter for international action; and the formulations preferred by the delegations who contributed to that report agree that sustainable development is key to dealing with the consequences of climate change – or at least it is for all those states that are likely to survive rising sea levels over the next few decades. As members of the Alliance of Small Island States made clear in their submissions to the Secretary General:

> The potentially affected Member States have put forward the view that, for them, no amount of sustainable development can protect against the security implications of climate change and that development itself becomes meaningless when there is no longer any sovereign territory with which it can be associated.
>
> *(UN 2009: 25–6)*

Sustainable development is irrelevant for states whose territories will be completely inundated; for them, existential security is of paramount importance.

For Republican politicians in Washington five years later, sustainable development was similarly irrelevant, albeit for very different reasons: US interests focus on the United States, not on the rest of the world. For the small island states, this narrow focus is precisely the locus of their security problem! However, while climate denial and the lobby campaigns of the fossil fuel industry are clear and present dangers to the small island states, those states have no military means with which to protect their territorial integrity. The UN too, it seems, is relatively powerless to act, at least as long as the threats to rich and powerful states continue to form the basis of the formulations of climate security. This absence of effective capabilities emphasizes the failure of international cooperation so far, certainly in terms of dealing with climate matters – or at least it does if the small island states and those most likely to suffer the consequences of accelerating climate change have any meaningful claim on global security.

Reframing climate change

The first half of this chapter focused on the use of the threat multiplier formulation as a warning of future dangers in the CNA Corporation report of 2007 and the UN Secretary General's report of 2009. Framing an issue in terms of threat multipliers was of much less importance to most UN member states than securing help with sustainable development to deal with climate change, something which, after all, developed states have largely imposed on poor countries. Correcting this geography is important to linking climate to security. Looking to peripheral states as the source of threats may work for US security planners, but it is not a credible argument for most of the rest of the world, where climate disruptions are caused indirectly by fossil fuel consumption in the rich parts of the global economy, not by local political instabilities (Dalby 2013).

Contrasting the 2007 CNA Corporation document with the 2014 CNA Military Advisory Board update emphasizes the shift from a frame of future dangers to one suggesting that climate is already a catalyst for security difficulties and hence an immediate priority rather than a future possibility. This shift in climate security discourse ought to facilitate a policy response to deal with climate change more urgently. However, as the arguments for the repudiation of climate science as the basis for US national security show, this attempt at securitization has substantially failed, at least for the moment, in Washington. Dismissing the changing environment as a matter of concern, as the repudiation of climate science does in the example discussed in this chapter, allows the rejection of the climate security frame and with it at least some policy attempts to constrain the production and use of fossil fuels.

The consequences of such a dismissal of science will clearly be severe, if one takes earth system science remotely seriously as a premise for understanding the world. A sustainable future in the Anthropocene looks very unlikely unless fossil fuel use is seriously constrained soon. Ironically, if fossil fuel production continues with its current abandon, and attempts to frame the likely climate consequences in terms of security continue to fail, the US armed forces' ability to respond to future disruptions is likely to be reduced because of funding constraints that will prevent planning and training to deal with disasters.

Language and politics are irredeemably interconnected. In the absence of an effective political strategy aimed at much more than mobilizing alarm as the basis for action (Cox 2010), reframing climate as a matter of national security, drawing upon authoritative science to issue warnings about what is coming, is not enough to move policy forward now in Washington, at least. Important though framing is in the case of climate change, the politics for dealing effectively with the issue will need more – very much more – than has so far been accomplished by contemporary attempts to securitize the issue.

References

Buzan, B., Waever, O., and deWilde, J. (1998) *Security: A New Framework for Analysis*, Boulder, CO: Lynne Rienner.

Campbell, K.M., Gulledge, J., McNeill, J.R., Podesta, J., Ogden, P., Fuerth, L., Woolsey, R.J., Lennon, A.T.J., Smith, J., Weitz, R., and Mix, D. (2007) *The Age of Consequences: The Foreign Policy and National Security Implications of Global Climate Change*, Washington, DC: Center for Strategic and International Studies and Center for a New American Security, http://csis.org/files/media/csis/pubs/071105_ageofconsequences.pdf, accessed 6 January 2015.

CNA Corporation (2007) *National Security and the Threat of Climate Change*, Alexandria, VA: CNA Corporation.

CNA Military Advisory Board (2014) *National Security and the Accelerating Risks of Climate Change*, Alexandria, VA: CNA Corporation.

Cox, J.R. (2010) "Beyond frames: Recovering the strategic in climate communication," *Environmental Communication*, 4(1): 122–33.

Dalby, S. (2009) *Security and Environmental Change*, Cambridge: Polity.

Dalby, S. (2013) "Climate change: New dimensions of environmental security," *RUSI Journal*, 158(3): 34–43.

Dalby, S. (2014) "Environmental geopolitics in the twenty-first century," *Alternatives: Global, Local, Political*, 39(1): 1–14.

Dalby, S. (forthcoming) "Geopolitics, ecology and Stephen Harper's reinvention of Canada," in H.G. Brauch, U. Oswald Spring, G. Grin, and J. Scheffran (eds) *Sustainability Transition and Sustainable Peace Handbook*, Heidelberg: Springer-Verlag.

Deudney, D. (1990) "The case against linking environmental degradation and national security," *Millennium*, 19(3): 461–76.

Dunlap, R. (2014) "Clarifying anti-reflexivity: Conservative opposition to impact science and scientific evidence," *Environmental Research Letters*, 9(2), http://iopscience.iop.org/1748-9326/9/2/021001, accessed 21 March 2015.

Fierke, K.M. (2007) *Critical Approaches to International Security*, Cambridge: Polity.

Finger, M. (1991) "The military, the nation state, and the environment," *Ecologist*, 21(5): 220–5.

Floyd, R. and Matthew, R. (eds) (2013) *Environmental Security: Approaches and Issues*, London and New York: Routledge.

German Advisory Council on Global Change (WBGU) (2008) *World in Transition: Climate Change as a Security Risk*, London: Earthscan.

Gleick, P. (2014) "Water, drought, climate change and conflict in Syria," *Weather, Climate and Society*, 6: 331–40.

Gutman, D, (2014) "McKinley amendment bars Defense funds for climate change," *Charleston Gazette*, 24 May, www.wvgazette.com/article/20140525/GZ01/140529501, accessed 20 January 2015.

Interagency Working Group on the Social Cost of Carbon (2013) *Technical Update of the Social Cost of Carbon for Regulatory Impact Analysis – Under Executive Order 12866*, Washington, DC: United States Government.

Intergovernmental Panel on Climate Change (IPCC) (2007) *Fourth Assessment Report*, Cambridge: Cambridge University Press.

Intergovernmental Panel on Climate Change (IPCC) (2014) *Fifth Assessment Report*, Cambridge: Cambridge University Press.

Jacques, P.J., Dunlap, R.E., and Freeman, M. (2008) "The organization of denial: Conservative think tanks and environmental skepticism," *Environmental Politics*, 17(3): 349–85.

Klein, N. (2014) *This Changes Everything: Capitalism vs. the Climate*, New York: Simon & Schuster.

McKinley, D. (2014) "Amendment to the Rules Committee print for H.R. 4435. Sec. 318. Prohibition on use of funds to implement certain climate change assessments and reports," 17 May.

Mathews, J.T. (1989) "Redefining security," *Foreign Affairs*, 68(2): 162–77.

Mayer, M. (2012) "Chaotic climate change and security," *International Political Sociology*, 6(2): 165–85.

Melillo, J.M., Richmond, T., and Yohe, G.W. (eds) (2014) *Climate Change Impacts in the United States: The Third National Climate Assessment*, Washington, DC: US Global Change Research Program.

O'Brien, K., Wolf, J., and Sygna, L. (eds) (2013) *The Changing Environment for Human Security: New Agendas for Research, Policy, and Action*. London: Earthscan.

Oels, A. (2012) "From 'securitization' of climate change to 'climatization' of the security field: Comparing three theoretical perspectives," in J. Scheffran, M. Brzoska, H.-G. Brauch, P.M. Link, and J. Schilling (eds) *Climate Change, Human Security and Violent Conflict: Challenges for Societal Stability*, Berlin: Springer-Verlag, pp. 185–205.

O'Lear, S., Briggs, C., and Denning, G.M. (2013) "Environmental security, military planning and civilian research: The case of water," *Environment*, 55(5): 3–12.

Pumphrey, C. (ed.) (2008) *Global Climate Change: National Security Implications*, Carlisle, PA: US Army War College, Strategic Studies Institute.

Scheffran, J., Brzoska, M., Brauch, H.-G., Link, P.M., and Schilling, J. (eds) (2012) *Climate Change, Human Security and Violent Conflict: Challenges for Societal Stability*, Berlin: Springer-Verlag.

Schwartz, P. and Randall, D. (2003) *An Abrupt Climate Change Scenario and Its Implications for United States National Security*, GBN, www.climate.org/PDF/clim_change_scenario.pdf, accessed 6 January 2015.

Sharman, A. (2014) "Mapping the climate sceptical blogosphere," *Global Environmental Change*, 26: 159–70.

Steffen, W., Persson, A., Deutsch, L., Zalasiewicz, J., Williams, M., Richardson, K., Crumley, C., Crutzen, P., Folke, C., Gordon, L., Molina, M., Ramanathan, V., Rockström, J., Scheffer, M., Schellnhuber, H.J., and Svedin, U. (2011) "The Anthropocene: From global change to planetary stewardship," *Ambio*, 40: 739–61.

United Nations (1994) *Human Development Report*, New York: United Nations Development Program.

United Nations (2009) *Climate Change and Its Possible Security Consequences*, New York: Report of the Secretary General A/64/350.

US Department of Defense (2014) *2014 Climate Change Adaptation Road Map*, Washington, DC: Department of Defense.

Webersik, C. (2010) *Climate Change and Security: A Gathering Storm of Global Challenges*, Santa Barbara, CA: Praeger.

Whitehead, M. (2014) *Environmental Transformations: A Geography of the Anthropocene*, London: Routledge.

Yergin, D. (2011) *The Quest: Energy, Security, and the Remaking of the Modern World*, New York: Penguin.

Zografos, C., Goulden, M.C., and Kallis, G. (2014) "Sources of human insecurity in the face of hydro-climatic change," *Global Environmental Change*, 29: 327–36.

7
GEOPOLITICS AND CLIMATE SCIENCE

The case of the missing embodied carbon

Shannon O'Lear

The Intergovernmental Panel on Climate Change (IPCC) is the institution that sets the parameters for international discussion on climate change. The IPCC was intended to capture scientific understanding of climate change and provide informed guidance to policy-makers. A widely held view of science is that it is distinct and separate from politics and that it therefore can provide objective guidance to policy. This view dates back to Vannevar Bush's report for President F.D. Roosevelt entitled *Science, the Endless Frontier* (Bush 1945). In that report, Bush promoted a dualistic vision of basic, research-based science, on the one hand, and applied product development, on the other hand: the "R" and "D" of "research and development" or the idea of pure, impartial science as necessarily distinct from political applications (see also Jasanoff and Wynne 1998). This binary view promotes science – rooted in certainty, fact, and truth – as playing an advisory role to values-driven policy (Price 1965) or, rather, as "speaking truth to power."

This chapter, however, looks at science as very much intertwined with social and political processes to the extent that these processes shape how science is portrayed to policy-makers. Crucially, it examines the taken-for-granted assumption that data assembled at the level of territorial states is the most appropriate mode of measurement for climate policy.

Every four or five years, the IPCC produces voluminous reports assessing published scientific work related to climate change.[1] These reports reflect the best and most comprehensive scientific understanding of climate change and its socio-economic impacts, and they are intended as guides to policy development. Each of the (now) three working groups of the IPCC compiles its own report, focused on:

1 "the physical scientific aspects of the climate system and climate change";
2 "the vulnerability of socio-economic and natural systems to climate change"; and

3 "options for mitigating climate change through limiting or preventing greenhouse gas emissions."

(IPCC 2014b)

The latest versions of these reports weigh in at well over 1,500 pages each. Each of the working groups distils its voluminous reports into an approximately 30-page "Summary for Policymakers" (SPM). Indeed, the IPCC was envisioned by its creators and members of scientific and environmental communities "to serve as a politically negotiated scientific baseline for binding international laws and policies" (Howe 2014: 153). What happens, though, when politics shapes not only how science is represented but also which science "counts"? This chapter considers the case of the most recent report of the IPCC, the *Fifth Assessment Report* (AR5). Controversial scientific information was pulled from a summary report for policy-makers because to include it would have challenged an established norm in the international response to climate change. This case offers a provocative example of how the structure of scientific representation within the IPCC, rather than encouraging a creative reframing of climate, instead perpetuates geopolitics as usual.

Framing climate with national science

Societies have always lived with climate, irrespective of whether they have called it that, and have developed ways to interpret and respond to lived experiences with long-term alterations in their physical environment. Scholars in the social sciences and humanities have long studied narratives of climate that individual societies have developed and passed on through generations (Hulme 2009; Carey 2012). There are many ways to interpret climate, but most of them, historically, have emerged in response to direct, lived experience with the physical environment.

The advent of computer technology, particularly in the 1960s, enabled the emergence of a completely different understanding of climate: the global climate. Only with the computational and simulation power of computers could such a thing as a global dataset be assembled from different sets of localized weather data (Edwards 2010). Satellite imagery and data collection further advanced the ability to study phenomena at the global scale that had not been possible before. Yet, at that time, the Cold War political rivalry between the United States and the Soviet Union was at its peak. The two states faced off in a struggle for global dominance politically and ideologically. The threat of nuclear war brought the concept of global freezing into popular nomenclature, and the coldest of Cold War years saw heightened interest in weather prediction and the potential for weather control with strategic military applications (Weart 2008). Funding from the US government promoted certain disciplines, methods, and forms of analysis over others (Weart 2013), and investing in "science" was a way to promote a particular political agenda. Think of the "space race," in which the United States and the

Soviet Union each tried to outdo the other in aeronautical programmes such as landing on the moon. These publicly visible applications of particular modes of science helped to promote national identity in an "us versus them" way. In the same way, the securitization and nationalization of science were extended to weather control and, eventually, climate science.

National priorities shaped funding priorities, and government-supported funding allowed the expansion of weather observation and data-collection capacities, the development of satellites and communication networks for processing data, and the establishment of university departments and programmes to train scientists in the advancing field of climate science. A significant part of this effort was to establish networks to support the development – and legitimacy – of climate science (Blok 2010). Institutions such as the National Science Foundation, the National Center for Atmospheric Research, and the National Oceanic and Atmospheric Administration were established, funded, and staffed to promote scientific advancement. Internationally, the World Meteorological Organization was also established as a branch of the United Nations to foster scientific collaboration, standardization, and exchange of data and training efforts among members. Of course, every "member" was a state.

The establishment of these research programmes and institutions served to generate information that could be used to inform policy decisions. State support of these efforts reflects, in part, the promotion of climate science as a national concern. Returning to the idea of creating "global" data through the use of technology and internationally connected research organizations (Edwards 2010), the nationalization of climate demonstrates a particular selection of "global" data. Once a large pattern such as circulation or another large-scale trend is identified from an assembled collection of data, governments or policy-makers can look at the implications of such patterns and trends for their own states. Such attempts to "see like a state" (Scott 1998) can involve "thin simplifications" of the complex interdependencies reflected in scientific understanding of climate. Mapping climate change onto the existing geopolitical map assumes that climate change may be understood and possibly managed on a territorial basis.

Because of the massive scale and complexity of climate, policy audiences have sought to stabilize an understanding of what is, in actuality, a dynamic and always incomplete field of study. Intertwined with a techno-scientific approach to climate is a postmodern view of environmental management. Climate change became an issue to be studied, measured, and controlled through accepted practices of cost–benefit analysis, technological development, and economic incentives – all established tools of the neoliberal state system (Glover 2006). Since the 1990s, in particular, the development of a global climate regime has framed climate as an issue to be managed or governed through market-based incentives and appropriate applications of technology (Oels 2005). As detailed in Glover's discussion (Chapter 2, this volume), the dominant approach to climate assumes that the modern state system has the capacity and reach to manage climate with the same prioritization of industrial, technological advancement and state-centric modernity that, arguably, brought about significant changes to climate in the first place.

By its very nature, capitalism's prioritization of competition generates waste, including environmental degradation (Peet et al. 2011). Environmental degradation and change, then, are parts of the neoliberal economic system. It is this established system, bound by a state-centred focus, that dominates much – but not all – of the current response to climate change. Global climate governance relies heavily on emissions trading, carbon-offset projects, and other features, which construct climate change as a market problem that can be addressed through neoliberal market strategies (Newell and Paterson 2010; see also Lohmann 2008). Such an approach reinforces the state system through regulatory norms and political protection of corporate interests. As established governing bodies, states certainly have a significant role to play in decision-making and policy support. However, embedded in the foundation of the state system is the principle of territorial sovereignty and non-intervention. What states do is their business, and this mutual recognition leaves little room for oversight in the interest of the greater good. This prioritization of the state as legitimate actor and appropriate unit of analysis falls squarely into what John Agnew (1994) has described as "the territorial trap." Such a focus on the territorial state often (wrongly) assumes that states are uniform actors with some form of homogeneous support emanating from within their boundaries. Focusing on states also misses the importance of things that move between and among states, such as ideas, information, people, investment, disease, and pollution. The territorial state reflects a form of jurisdictional sovereignty, but even the nature of how sovereignty is exercised, legitimized, or challenged varies widely from state to state. What is more, the notion of promoting or protecting national security reinforces the notions of boundaries to distinguish inside from outside and "us" from "them." The very notions and mechanics of territorial states raise sharp contradictions to the idea of environment-related collaboration (Deudney 1991; Dalby 2009).

This persistent state-centred "cancer of Westphalia" (Harris 2013) restricts how climate change is framed and limits the range of policy options and actors. Such a focus sidelines other actors and processes that could have significant impacts on how climate change unfolds. Industries, terms of international trade agreements, structures of finance (see Grove's discussion, Chapter 11, this volume), and economic pricing practices, for example, are avenues that could be pursued in the interest of altering negative human–environment interactions. In a state-centred approach to climate, however, these contributing elements often go unaddressed. What is more, other actors and interests wield significant, if only indirect, power. Indeed, "energy and fossil fuel-interested organisations and states" (de Coninck and Bäckstrand 2011: 368) have become dominant influences on the institutional landscape, thereby further reinforcing neoliberal economic values and state–corporate relationships.

The remainder of this chapter looks at the most recent report of the IPCC to illustrate how a selective portrayal of information helps to maintain rather than challenge climate change as geopolitics as usual. The argument here is that a different representation of human–environment interactions could be mobilized to reframe policy responses to climate change.

The IPCC, states, and the representation of science

When the IPCC was established in 1988, the scientists involved represented national laboratories and state agencies. Spencer Weart (2008: 153), a historian of climate science, has emphasized that, therefore, the IPCC is neither completely a scientific organization nor solely a political organization, but rather a "unique hybrid" of both. Yet, what is evident in the structure of the IPCC is that all states do not have an equal voice in decision-making processes. Leadership of the IPCC working groups, for example, was originally assigned based on the political importance of the represented countries (Howe 2014: 163), thus granting them greater potential to influence procedure. This was one way in which, even at its inception, the IPCC did not reflect "objective" science but had national interests embedded in its infrastructure. The state-centric approval structure of the IPCC has been critiqued more broadly elsewhere (Baer 2010; van der Sluijs et al. 2010; Wynne 2010; Hoffman 2011) since it has had a significant impact on how climate science is represented. For instance, a main scientific message of the IPCC has been that climate change is gradual and manageable within the current institutional setting, which downplays the significance of instabilities in atmospheric, oceanic, and ecological systems (Wynne 2010: 297). Embedded in that message is that the current arrangements of power do not need to change. Although that observation appears to be less true for the *Fifth Assessment Report*, which states the urgency of climate change more strongly than previous reports, there remains the concern that more attention should be paid to how and why science frames questions in the way that it does and how the science of climate change is interpreted and legitimized (Goeminne 2010). The role of government representatives in the production of documents for policy-makers reinforces this message of manageability and thereby constrains the range of response pathways offered in these documents. The very structure of the IPCC, then, inherently limits how much the IPCC might actually influence policy. The underlying geopolitical priorities shaping these outcomes are obscured precisely because they are hidden in plain view. A similar point is made by Naomi Klein (2014), who argues that among myriad other failures of the neoliberal state system, it has failed to address climate change in a meaningful way.

Relying too much on scientific problem-solving may obscure the moral imperative to act in the interest of human well-being and ecosystem sustainability (Lemos and Rood 2010). Uncertainty around global climate change comes in many forms and is only to be expected, given the complexities of climatic and human systems. Scientific uncertainty, of course, is part of the process. As Sarewitz (2004) argues, uncertainty drives forward scientific inquiry, but it is often misunderstood as a negative feature or shortcoming of science. Waiting for complete information would stall any decision-making, so at some point decisions must be made based on imperfect knowledge (and a decision not to act is still a decision). Here, it is important to recognize that "decisions about uncertainties are political (and ethical)" (Douglas 2006: 226). The richness and variety of scientific understanding could be valuably examined for the meaning and implications of divergent pieces of information. However, the state-centred approval process for the IPCC's "Summaries for Policymakers" (SPMs) has resulted in a selective representation of science. Rather than examine multiple approaches to the measurement of greenhouse gas (GHG) emissions, the following example highlights how some forms of evidence were, instead, removed from a policy document to perpetuate the status quo of policy arrangements and power relationships among states.

As noted in the previous section, the IPCC involves a vast array of scientists working on multiple dimensions of climate change and human–environment relationships. However, states, through government representatives, remain as gatekeepers to the SPMs generated by each of the three IPCC working groups. Government representatives have line-by-line approval rights over the information that is taken from each of the comprehensive scientific reports and included in the SPMs. Arguably, all of the science is still "there" in the larger reports, but there is little likelihood that policy-makers will go beyond the abbreviated reports that have been crafted for their use.

Following the release of the *Fifth Assessment Report*, some of the contributing authors to the Working Group III (WG III) SPM (IPCC 2014a) published papers in *Science*, a leading scholarly journal, detailing information in the full report that was excluded from the SPM for political reasons. The final draft of the WG III full report looks at research on change in GHG emissions from 1990 to 2010 across different regions of the world. A key finding is that "the most influential driving force for the emission growth has been the increase of per capita income" (Agrawala et al. 2014: 14). Additionally, compared to the 1990s, carbon dioxide emissions from fossil fuel use in the 2000s were higher in all regions except for transitional economies, and Asia showed the largest rate of increase in these emissions (Blanco et al. 2014: 10). According to Annex II of the WG III report, Asia includes non-OECD (Organization for Economic Cooperation and Development) countries of East Asia, South Asia, South-East Asia, and the Pacific. This group of countries includes China and India (Krey et al. 2014). Another chapter of the full report states that between 1990 and 2010, "CO_2 emissions from coal grew by 4.4 $GtCO_2$ in EAS [East Asia], which is equivalent to roughly half of the global net increase of CO_2 emissions from fossil fuel consumption" (Agrawala et al. 2014: 16). Looking specifically at China between 2002 and 2005, manufacturing for exports is the most significant driver of emissions, and the increased use of coal has also contributed to the increased rate of carbon dioxide emissions in the country (Agrawala et al. 2014: 17). Some of the contributing authors of the WG III SPM have emphasized that it is precisely these types of findings about certain categories of countries contributing more significantly to global emissions that were cut from the SPM (Victor et al. 2014). There are many ways to calculate emissions and to group countries, but the final report reflects options that do not challenge the current grouping of countries, methods of accounting for GHG emissions, or policy approaches. Instead, these contributing authors point out, what should happen is that richer, industrialized states should make a greater effort to cut emissions, and emerging, growing economies must accept that they are increasing their contributions to global GHG emissions. Key to this discussion is the way in which GHG emissions are measured.

One way to measure emissions is based on a country's territory. We see these kinds of data in use, for example, in tables comparing emissions generated within the territory of, say, China, the United States, and other territorial states. The emissions taken into consideration are calculated based on activities within each state's borders over a particular period of time. The United Nations Framework Convention on Climate Change (UNFCCC) utilizes IPCC reports and data, and originally grouped countries into categories on the basis of territorial emissions. Annex I countries are richer, industrialized countries as well as some countries in transition (and a subset of Annex I countries, known as the Annex B states, have made emission reduction commitments; see Verbruggen et al. 2011). All other countries are non-Annex I countries, which are granted more leniency in their ambitions for GHG emissions reduction because it is understood that trying to adjust their emissions could do significant harm to their economic growth. A key finding in the WG III report, however, is that emissions from middle-income countries, which comprise a subset of the non-Annex I countries, have been increasing at a more rapid rate than those of other countries, as noted above.

Currently, the IPCC accounting rules for GHG mitigation apply only to emissions within a territory or jurisdiction. Such territorial-based accounting does not consider embodied emissions reflected in the consumption of goods traded between countries:

> A challenge with a territorial-based emission accounting system ... is that connections between economies are not directly considered. In particular, international trade and investment flows provide a link between production and consumption in different countries. Ignoring these connections might result in a misleading analysis of the underlying driving forces of global, regional, and national emission trends and mitigation policies.
> *(Peters et al. 2011: 8903)*

"Embodied carbon" captures emissions embedded in consumption. It is a term used interchangeably with "carbon footprint," "embedded carbon," "carbon flows," and "virtual carbon." The embodied carbon associated with a functional unit, such as an individual, a household, a company, a government, or a country, may be understood as "the climate impact under a specified metric that considers all relevant emission sources, sinks, and storage in both consumption and production within the specified spatial and temporal system boundary"

(Peters 2010: 245). Consumption-based accounting of carbon emissions "corrects territorial emissions by adding emissions generated to produce imported goods and services, and subtracting those generated to produce exports" (Steinberger et al. 2012: 81). It should be noted that the emphasis on carbon provides a limited conceptualization, since many other compounds (such as nitrous oxide and sulphur dioxide) and processes (such as land use change) are implicated in climate change. However, this general approach enables a shift of attention from emission sources to drivers of emissions as well as emission transfers reflected in international trade (Peters 2010). In short, paying attention to embodied carbon helps us to see who is benefiting from the consumption of carbon emissions rather than anchoring those emissions to the territorial place where they are generated. Since carbon dioxide is a uniformly mixed pollutant and does not stay in one place, assigning it to a particular state's territory is an economic accounting manoeuvre that reinforces the state-centred geopolitical system. Accounting measures could just as well transfer responsibility for emissions to places or activities of consumption driving the emissions. Since "consumption based emissions are more closely associated with GDP than are territorial emissions" (Fleurbaey et al. 2014: 30), they would seem to offer a more just means of accounting for carbon dioxide emissions since responsibility for the emissions would more closely reflect their benefits rather than compound their negative impacts. Chapter 4 of the full WG III report also considers risks associated with embodied carbon accounting, such as impacts on economic competitiveness and ecological sustainability trade-offs.

A focus on embodied carbon shows trends that are not captured in the territorial-based accounting of emissions. One group of researchers has noted that:

> the global emissions associated with consumption in many developed countries have increased with a large share of the emissions originating in developing countries. This finding may benefit economic growth in developing countries, but the increased emissions could also make future mitigation more costly in the developing countries. In addition, we find that the emission transfers via international trade often exceed the emission reductions in the developed countries. Consequently, increased consumption in the Annex B countries has caused an increase in global emissions contrary to the territorial emission statistics reported to the UNFCCC.
> *(Peters et al. 2011: 8907)*

Here, it becomes quite clear that the way in which emissions are measured matters not only for international efforts to reduce emissions, but also for particular countries that import or export a significant level of emissions. A consumption-based approach highlights an injustice embedded in territorial accounting of carbon emissions, namely, "Adjusting for the transfer of emissions through trade ... shows that further socio-economic benefits are accruing to carbon-importing rather than carbon-exporting countries" (Steinberger et al. 2012: 81). In other words, richer countries tend to enjoy the benefits of imported goods without the costs of emissions generated by the production of those goods in their country of origin. If we look again at the specific findings of the full report described earlier in this section, it is clear that accounting for GHG emissions on the basis of territory hides how consumption in richer countries in North America, Western Europe, Japan, Australia, and New Zealand (OECD countries) contributes to GHG emissions through carbon footprints and trade that tie this consumption to manufacturing in emerging economies. Also, if policy schemes regrouped countries to reflect their changing economic structure, then China, for instance, would likely be expected to take greater responsibility for its emissions as its wealth per capita increases.

Not only is the overall level of embodied or virtual carbon increasing at a global level, it is increasing unevenly. China, Russia, South Africa, and other middle-income countries are becoming increasingly significant net exporters of embodied carbon while the main net importers are the European Union, the United States, and Japan. Additionally, when disaggregated to the sector level within a country, there is significant variation in terms of carbon intensities across economic sectors. "Put another way . . . looking only at the overall carbon balance for a country masks a considerable amount of ebb and flow" (Atkinson et al. 2011: 565). Therefore, establishing a flat tax or tariff for carbon traded across borders could potentially do more harm to particular economic sectors within a country. Specifically, if richer countries tax incoming, embodied carbon associated with goods originating in less economically developed countries, the volume and composition of trade will likely change and have negative impacts on developing countries. However, multilateral trade agreements, such as those incorporated into the Montreal Protocol, could be an effective means towards climate mitigation if they are introduced with a generous grace period and accompanied by "effective finance and technology transfer mechanisms" (Zhang 2009: 5111). From an overview of databases and methods used in the growing number of studies on embodied emissions in trade, the volume of carbon dioxide embodied in global trade appears to be increasing, even though there remains considerable uncertainty in the estimations of embodied emissions at the national level. Even with these state-level disparities, "explicitly incorporating consumption based principles can, in theory, improve fairness of outcomes in terms of the distribution of responsibility across producers and consumers" (Sato 2013: 851). Utilizing consumption-based accounting of emissions as a shadow indicator could be helpful towards the development of more meaningful strategies for carbon dioxide mitigation.

Contributing authors to the WG III report included information on consumption-based emissions and embodied carbon in the full *Fifth Assessment Report* since the considerable amount of research in these areas provides an innovative way to understand GHG emissions. A group of these contributing authors commented:

> Studies on effects of trade and globalization have tracked emissions "embodied" in products that are traded across borders. For the first time, [the] IPCC presented adjusted emission statistics showing how territorial and consumption-based accounting systems lead to very different pictures. But because governments couldn't agree on how to group countries, all WG III findings about embodied emissions were cut from the SPM.
>
> *(Victor et al. 2014: 35)*

Measuring emissions to include not only emissions generated within a state's territory but also emissions associated with commodity chains of consumption challenges the current grouping of states into Annex I and non-Annex I countries.

Adjusting the groupings so that the trend in increasing emissions among middle-income countries is visible, however, would attract attention to the fact that perhaps those states should be held to more rigid standards for emissions reductions. That, of course, is not in their economic interest. This conflict over how and which versions of climate science, accounting, and social science are captured and presented to policy-makers raises questions about the relationship between the representation of science and state government interests. This example of the political approval of "acceptable" science to be shared with policy-makers demonstrates that global climate change is being selectively arranged to fit and perpetuate the familiar geopolitical map. Instead of adopting innovative approaches to understanding and addressing climate change, the priority, evidently, is to present science selectively and in a way that promotes the status quo. Policy aims to promote the interests of a given state, so, arguably, it is the policy process itself that reinforces – if not causes – political gridlock on climate change. Hence, meaningful responses to climate change may have to come from sources and agents other than states, and reframing climate change discussions so that they are not constrained by territorial states offers potential ways forward.

Beyond a geopolitical frame

The stated intention/purpose of the IPCC is to assess the myriad scientific approaches to understanding climate change and to reflect the collaborative and dynamic nature of science as a complex process. Yet, in this chapter I have explored an example of direct political influence on the way in which the breadth and depth of scientific understanding is funnelled selectively to policy-makers. This situation exemplifies what Erik Swyngedouw (2010, 2013) has termed a "post-political process," in that discussion is framed in a limited way to the advantage of certain groups, and consensus is constructed without the possibility of reconceptualizing the issue. Institutions such as the IPCC and the UNFCCC can entrench certain arrangements of power and promote a particular view of how a problem is understood. In the case of the WG III SPM, government representatives vetoed the inclusion of a set of scientific evidence – consumption-based emission measures – because that information would have challenged policy arrangements that currently grant middle-income states greater flexibility and less responsibility to reduce their emissions while expanding their economies. Instead of engaging with a genuine willingness to generate alternative options for compromise and negotiation, the process appears to be "post-political" because the outcome is predetermined by the rigid positions taken by the more powerful actors. The IPCC's representation of climate change to policy-makers remains safely within the confines of established power relations such that change is limited, predefined, and non-threatening to current arrangements of power.

It would seem that a new form of politics is needed if climate change is going to be reframed as an important situation to address proactively. Politics as usual, based on geopolitical imbalances of power that perpetuate many of the contributing factors of climate change, is not working. Stellan Vinthagen (2013: 156) has observed that "climate change is not, in a fundamental sense, possible to understand from a natural science perspective, but demands an analysis of political power relations and our existing global political economy, capitalism." He points out that the IPCC sets the dominant world narrative on the science of climate change and documents current scientific understanding of climate change. We already know a great deal about technical and behavioural changes that could (and probably should) be made to Western lifestyles. What is needed, Vinthagen argues, is a "Social Science Panel on Climate Change" to help us to figure out how to change our political economy in support of our claimed interest to save ourselves and the planet. Taking the current power structures into consideration, such a panel could apply skills and knowledge towards the development of alternative political and economic models and generate strategies to "resist anti-human and anti-ecological tendencies of world capitalism and nation-states" (Vinthagen 2013: 172).

From a different but complementary vantage point, Matthew J. Hoffman (2011) has investigated how organizations, sub-state government agencies, universities, businesses, and cities are working beyond the toolbox of the UNFCCC's Kyoto Protocol to enact voluntary and experimental responses to climate change. He refers to these innovative initiatives as experimental governance activity emerging out of disillusionment with the ineffectiveness of multilateral agreements regarding climate change. When we focus solely on state-centric, multilateral treaty-making to capture a global response to climate change, all of this momentum is overlooked. Hoffman is not suggesting that there is no role for states and international organizations such as the IPCC. Instead, he makes the case that multilateral treaties could be crafted "to facilitate scaling up, linking, and furthering the dynamics bubbling up from below" (Hoffman 2011: 161).

In the same way, researchers working on embodied carbon are not necessarily suggesting a complete overhaul in accounting for emissions: "Although we believe that territorial emission statistics should still remain central to climate policy, our results show a need for a regular monitoring, verification, and reporting of emission transfers via international trade" (Peters et al. 2011: 8907). Yet, for now, this suggested reframing of how individuals, households, businesses, cities, governments, and countries understand and respond to information about emissions is discussed only in WG III's voluminous full report. Policymakers who rely on the SPM of that full report will not have the benefit of this alternative view.

The IPCC process and other dominant debates and discussions about climate emphasize an interpretation of climate anchored to the geopolitical map. What if we could break out of this "territorial trap" and take a wider view of the meaning of climate change? Vannevar Bush's view of pure, objective science as a guide to policy does not hold for climate. Generating scientific findings does not lead directly to "correct" policy. This chapter has referred to multiple interpretations and accounting schemes for GHG emissions and different ways to group countries for policy development purposes. What if multiple dimensions of interpretation of climate were brought to the fore of humanity's response to climate change? Responding to climate change could become more creative, more multifaceted, more complex, but also more effective. Perhaps appreciating complexity is key since, clearly, a simplified, top-down, state-centred approach is not working. Instead of fixing carbon and other emissions to the state territories where they are generated, a focus on embodied carbon could be more spatially aligned with the dynamic flow of raw materials, commodities, investment, and pollution between states and regions and across borders. Indeed, an ecological geopolitics would not necessarily follow the borders and priorities of states. Accountability need not be associated exclusively with territorial states but could be associated with trade agreements, corporations, financial arrangements, and consumption choices. Breaking away from the familiar geopolitical spatialization of climate is likely to be critical to salvaging what remains of the ecological systems that support our existence and themselves span political borders.

Climate change is not like textbook collective action problems, and it also challenges a paradigm of individual responsibility (Jamieson 2014). It is unlike anything we have encountered. Industrialization, globalization, and globe-sprawling neoliberal capitalism are deeply implicated in the creation of this situation, but it is well beyond the scope of individuals and individual societies to address in any comprehensive way. We cannot expect any single version of science to present us with a clear solution. However, we can likely expect researchers in the physical sciences, social sciences, and humanities to continue to engage with climate-related issues and other aspects of human–environment relationships. If embodied carbon were to be counted, what kinds of institutions would need to be developed to hold consumers, industries, and states accountable for the impacts of commodity chains? Crafting summaries for policy-makers, who speak for the state, may be a failed strategy from the start. What might a summary for concerned consumers look like? Or for pension-fund managers? Or for educational institutions, energy system designers, sectors of the transport industry, or other groups or actors identified on the basis of non-geopolitical frames?

The roots of climate science may lie in the physical sciences, but social scientists and scholars in the humanities are uniquely positioned to challenge dominant modes of understanding and the trap of framing climate within the state system. It is likely to be up to them to generate innovative, provocative ways for us to think through our values and priorities and, hopefully, act upon them.

Note

1 According to the InterAcademy Council's *Review of the IPCC*, the IPCC also draws from "gray literature," including reports, datasets, bulletins, conference papers and presentations, which vary in their level of peer review and quality (IAC 2010: 16). IPCC working groups focus less on physical science and more on socio-economic issues and draw more information from these sources, as noted by Gleditsch (2012: 4).

References

Agnew, J. (1994) "The territorial trap: The geographical assumptions of international relations theory," *Review of International Political Economy*, 1: 53–80.

Agrawala, S., Klasen, S., Acosta Moreno, R., Barreto, L., Cottier, T., Guan, D., Gutierrez-Espeleta, E.E., Gámez Vázquez, A.E., Jiang, L., Kim, Y.G., Lewis, J., Messouli, M., Rauscher, M., Uddin, N., and Venables, A. (2014) "Regional development and cooperation," in O. Edenhofer, R. Pichs-Madruga, Y. Sonoka, E. Farahani, S. Kadner, K. Seyboth, A. Adler, I. Baum, S. Brunner, P. Eickemeier, B. Kriemann, J. Savolainen, S. Schlömer, C. von Stechow, T. Zwickel, and J.C. Minx (eds) *Climate Change 2014: Mitigation of Climate Change: Contribution of Working Group III to the Fifth Assessment Report of the Intergovernmental Panel on Climate Change*, Cambridge and New York: Cambridge University Press, pp. 1083–140.

Atkinson, G., Hamilton, K., Ruta, G., and Van Der Mensbrugghe, D. (2011) "Trade in 'virtual carbon': Empirical results and implications for policy," *Global Environmental Change*, 21: 563–74.

Baer, P. (2010) "The situation of the most vulnerable countries after Copenhagen," *Ethics, Place and Environment*, 13: 223–8.

Blanco, G., Gerlagh, R., Suh, S., Barrett, J., de Coninck, H.C., Diaz Morejon, C.F., Mathur, R., Nakicenovic, N., Ofosu Ahenkora, A., Pan, J., Pathak, H., Rice, J., Richels, R., Smith, S.J., Stern, D.I., Toth, F.L., and Zhou, P. (2014) "Drivers, trends and mitigation," in O. Edenhofer, R. Pichs-Madruga, Y. Sonoka, E. Farahani, S. Kadner, K. Seyboth, A. Adler, I. Baum, S. Brunner, P. Eickemeier, B. Kriemann, J. Savolainen, S. Schlömer, C. von Stechow, T. Zwickel, and J.C. Minx (eds) *Climate Change 2014: Mitigation of Climate Change: Contribution of Working Group III to the Fifth Assessment Report of the Intergovernmental Panel on Climate Change*, Cambridge and New York: Cambridge University Press, pp. 351–411.

Blok, A. (2010) "Topologies of climate change: Actor-network theory, relational-scalar analytics, and carbon-market overflows," *Environment and Planning D: Society and Space*, 28: 896.

Bush, V. (1945) *Science, the Endless Frontier*, Washington, DC: US Government Printing Office.

Carey, M. (2012) "Climate and history: A critical review of historical climatology and climate change historiography," *Wiley Interdisciplinary Reviews: Climate Change*, 3: 233–49.

Dalby, S. (2009) *Security and Environmental Change*, Cambridge: Polity.
de Coninck, H. and Bäckstrand, K. (2011) "An international relations perspective on the global politics of carbon dioxide capture and storage," *Global Environmental Change*, 21: 368–78.
Deudney, D. (1991) "Environment and security: Muddled thinking," *Bulletin of the Atomic Scientists*, 47: 22–9.
Douglas, H. (2006) "Bullshit at the interface of science and policy: Global warming, toxic substances, and other pesky problems," in G.L. Hardcastle and G.A. Reisch (eds) *Bullshit and Philosophy: Guaranteed to Get Perfect Results Every Time*, Chicago, IL: Open Court, pp. 215–28.
Edwards, P.N. (2010) *A Vast Machine: Computer Models, Climate Data, and the Politics of Global Warming*, Cambridge, MA: MIT Press.
Fleurbaey, M., Kartha, S., Bolwig, S., Chee, Y.L., Chen, Y., Corbera, E., Lecocq, F., Lutz, W., Muylaert, M.S., Norgaard, R.B., Okereke, C., and Sagar, A.D. (2014) "Sustainable development and equity," in O. Edenhofer, R. Pichs-Madruga, Y. Sonoka, E. Farahani, S. Kadner, K. Seyboth, A. Adler, I. Baum, S. Brunner, P. Eickemeier, B. Kriemann, J. Savolainen, S. Schlömer, C. von Stechow, T. Zwickel, and J.C. Minx (eds) *Climate Change 2014: Mitigation of Climate Change: Contribution of Working Group III to the Fifth Assessment Report of the Intergovernmental Panel on Climate Change*, Cambridge and New York: Cambridge University Press, pp. 283–350.
Gleditsch, N.P. (2012) "Whither the weather? Climate change and conflict," *Journal of Peace Research*, 49: 3–9.
Glover, L. (2006) *Postmodern Climate Change*, New York: Routledge.
Goeminne, G. (2010) "Climate policy is dead, long live climate politics!," *Ethics, Place and Environment*, 13: 207–14.
Harris, P.G. (2013) *What's Wrong with Climate Politics and How to Fix it*, Cambridge: Polity.
Hoffman, M. (2011) *Climate Governance at the Crossroads: Experimenting with a Global Response after Kyoto*, Oxford: Oxford University Press.
Howe, J.P. (2014) *Behind the Curve: Science and the Politics of Global Warming*, Seattle, WA: University of Washington Press.
Hulme, M. (2009) *Why We Disagree about Climate Change*, Cambridge: Cambridge University Press.
IAC (2010) *Climate Change Assessments: Review of the Processes & Procedures of the IPCC*, Amsterdam: InterAcademy Council, http://reviewipcc.interacademycouncil.net/report.html, accessed 4 November 2014.
Intergovernmental Panel on Climate Change (IPCC) (2014a) "Summary for Policymakers," in O. Edenhofer, R. Pichs-Madruga, Y. Sonoka, E. Farahani, S. Kadner, K. Seyboth, A. Adler, I. Baum, S. Brunner, P. Eickemeier, B. Kriemann, J. Savolainen, S. Schlömer, C. von Stechow, T. Zwickel, and J.C. Minx (eds) *Climate Change 2014: Mitigation of Climate Change: Contribution of Working Group III to the Fifth Assessment Report of the Intergovernmental Panel on Climate Change*, Cambridge and New York: Cambridge University Press, http://report.mitigation2014.org/spm/ipcc_wg3_ar5_summary-for-policymakers_approved.pdf, accessed 24 August 2014.
Intergovernmental Panel on Climate Change (IPCC) (2014b) "Working groups/task force," www.ipcc.ch/working_groups/working_groups.shtml, accessed 23 October 2014.
Jamieson, D. (2014) *Reason in a Dark Time: Why the Struggle against Climate Change Failed – and What It Means for Our Future*, Oxford: Oxford University Press.
Jasanoff, S. and Wynne, B. (1998) "Science and decisionmaking," in S. Rayner and E.L. Malone (eds) *Human Choice and Climate Change*, Volume 1: *The Societal Framework*, Washington, DC: Battelle Press, pp. 1–87.

Klein, N. (2014) *This Changes Everything: Capitalism vs. the Climate*, New York: Simon & Schuster.

Krey, V., Masera, O., Blanford, G., Bruckner, T., Cooke, R., Fisher-Vanden, K., Haberl, H., Hertwich, E., Kriegler, E., Mueller, D., Paltsev, S., Price, L., Schlömer, S., Ürge-Vorsatz, D., van Vuuren, D.P., and Zwickel, T. (2014) "Annex II: Metrics & Methodology," in O. Edenhofer, R. Pichs-Madruga, Y. Sonoka, E. Farahani, S. Kadner, K. Seyboth, A. Adler, I. Baum, S. Brunner, P. Eickemeier, B. Kriemann, J. Savolainen, S. Schlömer, C. von Stechow, T. Zwickel, and J.C. Minx (eds) *Climate Change 2014: Mitigation of Climate Change: Contribution of Working Group III to the Fifth Assessment Report of the Intergovernmental Panel on Climate Change*, Cambridge and New York: Cambridge University Press, pp. 1281–328.

Lemos, M.C. and Rood, R.B. (2010) "Climate projections and their impact on policy and practice," *Wiley Interdisciplinary Reviews: Climate Change*, 1: 670–82.

Lohmann, L. (2008) "Carbon trading, climate justice and the production of ignorance: Ten examples," *Development*, 51: 359–65.

Newell, P. and Paterson, M. (2010) "The politics of the carbon economy," in M.T. Boykoff (ed.) *The Politics of Climate Change: A Survey*, New York: Routledge, pp. 77–95.

Oels, A. (2005) "Rendering climate change governable: From biopower to advanced liberal government?," *Journal of Environmental Policy and Planning*, 7: 185–207.

Peet, R., Robbins, P., and Watts, M.J. (2011) "Global nature," in R. Peet, P. Robbins, and M.J. Watts (eds) *Global Political Ecology*, New York: Routledge, pp. 1–48.

Peters, G.P. (2010) "Carbon footprints and embodied carbon at multiple scales," *Current Opinion in Environmental Sustainability*, 2: 245–50.

Peters, G.P., Minx, J.C., Weber, C.L., and Edenhofer, O. (2011) "Growth in emission transfer via international trade from 1990 to 2008," *Proceedings of the National Academy of Sciences of the United States of America*, 108: 8903–8.

Price, D.K. (1965) *The Scientific Estate*, Cambridge, MA: The Belknap Press of Harvard University Press.

Sarewitz, D. (2004) "How science makes environmental controversies worse," *Environmental Science and Policy*, 7: 385–403.

Sato, M. (2013) "Embodied carbon in trade: A survey of the empirical literature," *Journal of Economic Surveys*, 28: 831–61.

Scott, J.C. (1998) *Seeing Like a State: How Certain Schemes to Improve the Human Condition Have Failed*, New Haven, CT: Yale University Press.

Steinberger, J.K., Roberts, J.T., Peters, G.P., and Baiocchi, G. (2012) "Pathways of human development and carbon emissions embodied in trade," *Nature Climate Change*, 2: 81–5.

Swyngedouw, E. (2010) "Apocalypse forever? Post-political populism and the spectre of climate change," *Theory, Culture & Society*, 27: 213–32.

Swyngedouw, E. (2013) "The non-political politics of climate change," *ACME: An International E-Journal for Critical Geographies*, 12: 1–8.

van der Sluijs, J.P., van Est, R., and Riphagen, M. (2010) "Beyond consensus: Reflections from a democratic perspective on the interaction between climate politics and science," *Current Opinion in Environmental Sustainability*, 2: 409–15.

Verbruggen, A., Moomaw, W., and Nyboer, J. (2011) "Annex I: Glossary, Acronyms, Chemical Symbols and Prefixes," in O. Edenhofer, R. Pichs-Madruga, Y. Sokona, K. Seyboth, P. Matschoss, S. Kadner, T. Zwickel, P. Eickemeier, G. Hansen, S. Schlömer, and C. von Stechow (eds) *IPCC Special Report on Renewable Energy Sources and Climate Change Mitigation*, Cambridge and New York: Cambridge University Press, www.ipcc.ch/pdf/special-reports/srren/SRREN_Annex_Glossary.pdf, accessed 31 October 2014.

Victor, D.G., Gerlagh, R., and Baiocchi, G. (2014) "Getting serious about categorizing countries," *Science*, 345: 34–6.

Vinthagen, S. (2013) "Ten theses on why we need a 'Social Science Panel on Climate Change,'" *ACME: An International E-Journal for Critical Geographies*, 12: 155–76.

Weart, S.R. (2008) *The Discovery of Global Warming*, revised and expanded edition, Cambridge, MA: Harvard University Press.

Weart, S. (2013) "Rise of interdisciplinary research on climate," *Proceedings of the National Academy of Sciences*, 110: 3657–64.

Wynne, B. (2010) "Strange weather, again: Climate science as political art," *Theory, Culture & Society*, 27: 289–305.

Zhang, Z.X. (2009) "Multilateral trade measures in a post-2012 climate change regime? What can be taken from the Montreal Protocol and the WTO?," *Energy Policy*, 37: 5105–12.

8
TECHNOLOGY AND POLITICS IN THE ANTHROPOCENE

Visions of "solar radiation management"

Thilo Wiertz

Introduction

In September 2013, the Intergovernmental Panel on Climate Change (IPCC) presented its latest report on the "physical science basis" of climate change (IPCC 2013). The report primarily underpins previous assessments and leaves no reasonable doubt that anthropogenic emissions of greenhouse gases (GHG) are leading to a rise in global average temperatures, and that strong mitigation efforts are required to limit temperature rise to two degrees Celsius (or even less) within this century. It is the last bullet point of the "Summary for Policymakers" that points to perhaps the most striking novelty within the report; for the first time in its history, the IPCC provides an in-depth review of research on "geoengineering," defined as "[m]ethods that aim to deliberately alter the climate system to counter climate change" (IPCC 2013: 29). Concerns about the lack of progress in international climate negotiations have led some scientists to consider technological interventions into the climate system as a potential "Plan B" to counteract some of the effects of global warming (Royal Society 2009). Two geoengineering approaches are usually distinguished: sequestering carbon dioxide from the ambient air, so-called "carbon dioxide removal"; and increasing the planetary albedo – that is, the reflectivity of the planet – in order to reduce energy uptake and thereby temperature rise, so-called "solar radiation management."

Over the last few years, the interest in geoengineering within scientific communities, but also in policy circles, has increased significantly. In particular, solar radiation management (SRM) has been discussed controversially since Nobel Laureate Paul Crutzen raised the topic in an editorial essay in the journal *Climatic Change* (Crutzen 2006). Now, a wide range of ideas on counteracting global temperature rise by shielding sunlight from the earth populates the academic literature, and while some proposals, such as placing mirrors in space, clearly fall into the category of science fiction, others are considered technically less challenging. Among the more plausible proposals are, for example, injections of sulphur dioxide into the atmosphere and brightening marine clouds by introducing sea-salt particles.

The discussion of SRM is controversial particularly due to the large uncertainties and risks that such interventions would entail. SRM would only imperfectly counteract rising GHG concentrations, would likely lead to shifts (for example, in rainfall patterns), and would not stop ocean acidification. Furthermore, a sudden cessation of any SRM intervention would cause a rapid rise in global temperatures (IPCC 2013). Consequently, rather than replacing the need for emissions reduction, SRM would lead to a spatial and temporal redistribution of climate risks, raising the question of how decisions over such an intervention would be made. Concerns are fuelled by the proposition that SRM technologies could be much cheaper than mitigating GHG and could be implemented by a single state or even individual actor (Barrett 2008; Victor et al. 2009), and many fear that the prospect of a Plan B would become an excuse to delay urgent emission cuts further (Preston 2013). Certainly, the vision of a cheap and powerful tool to manipulate the global environment has the potential to unsettle some of the foundations of climate politics.

The discussion about the potential benefits and risks of SRM entails a fundamental shift in framing human–nature relations. While much of the climate change discussion has identified the influence of human activities on the global environment as a threat to environmental (and social) stability, in geoengineering this influence is reframed as an opportunity – a potential "lever" that may be used purposefully to direct and manage environmental processes on a planetary scale. If the focus of climate politics thus far has been on limiting anthropogenic interferences with the atmosphere, geoengineering raises the question of the extent to which humans can and should actively take control. This move from "withdrawal" to "intervention" parallels a paradigmatic shift in scientific knowledge production that has taken place in the fields of climate and earth system science. These two, largely overlapping fields have conceptualized the world as a globally coupled system, a "machine" of interacting components, and provide the conceptual background against which different ideas for geoengineering have been developed.

Drawing on insights from governmentality studies to focus on the role of scientific knowledge in conceptualizing the relations between humans, technology, and environmental change, this chapter examines how the idea of global climate interventions by geoengineering has emerged as a rationality for governing climate change from the context of earth system science and climate modelling. Foucault's work on governmentality points to the intricate relationship between the logics that characterize the production of knowledge about a specific domain and the practices, methods, or strategies of governing that domain (Rose and Miller 1992; Foucault 2007). An extension of his work on questions of "green governmentality" asks how knowledge about "nature" and its relation to human activities is produced and how this privileges certain ways of governing over others (Rutherford 2007). This chapter argues that an earth system science approach to human–environment relations often considers the politics of these relations as a predominantly global and – in the case of geoengineering, literally – technical matter, envisioning an "ecological geopolitics" that positions experts at the forefront of solving the "Anthropocene crisis." Computer simulations are an essential epistemic tool in this task as they produce calculable visions of the future that allow for notions of climate control and optimization.

The next section outlines how SRM has emerged from the context of earth system science and is related to earlier visions of global environmental management. Climate models have been central to the development of an earth systems perspective, and the third section argues that it is through computer simulation studies that SRM could become a matter of rational scientific calculus. The fourth section critically reflects on the limitations that characterize climate model-based visions of the future, particularly when trying to consider the social and political implications of SRM. The conclusion proposes the strengthening of interdisciplinary engagements that do not take the scientific framing of geoengineering as a starting point.

Technology and politics in an earth systems perspective

Science has a key role within the discourse on geoengineering. This is not surprising, given the key uncertainties around the ecological changes that would result from technical interventions in the climate system. However, the political relevance of science exceeds that of an informant about physical realities; research on knowledge–power relations in environmental discourses emphasizes that scientific knowledge production has implications for the way in which relations between people and their environments are understood and organized (Escobar 1996; Braun and Wainwright 2001). Stephanie Rutherford (2007: 298) points out that ecology and the earth sciences "have become fundamental to the production of regimes of governmentality that create the conditions of possibility to speak about nature as something in desperate need of governing by particularly located experts." This raises the question of how science is tied into discourses that privilege certain ways of governing human–nature relations over others. While the discussion on geoengineering has become increasingly diverse over recent years and cannot be subsumed under a single logic, the idea of technological climate interventions has strong roots within the contexts of climate and earth system science. The systems perspective developed in these fields has also become a powerful basis for propositions about how to govern human–environment interactions more efficiently.

The world seen from space has perhaps become the most powerful icon of international environmental discourses. The photographs taken by the Apollo missions during the Cold War era have been interpreted as symbolic of the unity and fragility of humanity's "home planet" and of the need to cooperate across cultural and political boundaries (Cosgrove 2006). At about the same time, another method of looking at the earth from a distance became increasingly powerful, albeit less visible to a wider public audience. Cold War investments in science and computational power facilitated the development of numerical simulation models of atmospheric processes, not least driven by a military interest in improved weather forecasts as well as methods to manipulate the weather (Fleming 2006). Alongside these developments emerged an increasing interest in simulating atmospheric processes on a global scale (Weart 2010; Edwards 2011). This interest became politically relevant when, in the early 1980s, scientists began to warn about the danger of a "nuclear winter," an ice age that could result from a nuclear confrontation between the United States and the Soviet Union. The dust that would rise after a series of nuclear detonations would reflect considerable amounts of sunlight, and with the support of global climate models it had become possible to project the resulting changes in temperatures around the world (Crutzen and Birks 1982; Perry 1985). In parallel, scientists started to use climate models to examine the climate effects from an increase in atmospheric carbon dioxide concentrations, pointing to the possibility of significant global warming caused by the burning of fossil fuels (Weart 2010). While the Apollo images underlined the cosmic presence of the world – a singular timeless entity floating in space – climate models presented the world as a system of complex interacting processes. The power of this "virtual" perspective derived from its potential to study the *changes* that would result from human activities and to create visions of hypothetical climatic and environmental futures.

Following discussions about nuclear winter and anthropogenic global warming, the systems approach to global climate and environmental changes became increasingly institutionalized. In 1972, the International Institute for Applied Systems Analysis (IIASA) was founded to enable scientific cooperation across the Iron Curtain and in 1987 several researchers, Paul Crutzen among them, successfully argued for the establishment of an international research programme, the International Geosphere Biosphere Programme (IGBP). By reaching out to other natural scientific disciplines, the IGBP has been particularly active in extending the systems approach from climate science into an integrated "earth systems science" framework that studies the relations between different "spheres" of the earth system: for example, interactions between the atmosphere, hydrosphere, and biosphere. Increasingly, this also included a consideration of humans as an additional component, or sphere, of the coupled earth system and motivated the foundation of an Earth System Science Partnership that combined different research frameworks under the auspices of the International Council for Science (Lovbrand et al. 2009). With the foundation of the IPCC in 1988, and its regular reports on the current state of climate research – relying strongly, albeit not exclusively, on results from computer modelling – the earth systems perspective became ever more intertwined with international environmental politics (Miller 2004).

The institutionalization of an integrated earth systems science approach has been motivated by concerns about the environmental changes that human activities would cause. Beyond establishing a novel approach to the study of ecological systems, this has also inspired propositions about how to manage the global environment more effectively as part of a "normative agenda of improving or saving the world" (von Storch 2004: 245). Although geoengineering has played a marginal role within such discussions, at least until recently, the idea has roots in the same institutional and discursive context. The first use of the term "geoengineering" in relation to climate change is usually attributed to Cesare Marchetti, who, in 1976, wrote a research memorandum published by the IIASA (Marchetti 1976). The Italian physicist suggested that carbon dioxide could be stored as biomass that could then be dumped into the ocean to remove it from the global carbon cycle. He saw this approach "in the spirit of geoengineering, which is a kind of 'system synthesis' where solutions to global problems are attempted from a global view" (Marchetti 1976: iii). This connection between global problem definitions and equally global solutions is not particular to the idea of geoengineering, but characterizes the perspective of earth system science more generally and has often been related to notions of "efficiency," "management," and "optimization." In 1999, Philip Newton saw a "manual for planetary management" emerging from the IGBP and asked how "scientists [might] increase the chances of effective management of global change" and how improved scientific understanding might be used "to optimize the role of policy-makers" (Newton 1999: 399). In the same year, John Schellnhuber (1999: C19) prominently asserted that earth system analysis was bringing about a "second Copernican revolution" that was attempting "to understand the 'earth system' as a whole and to develop, on this cognitive basis, concepts for global environmental management." He posited that geoengineering could provide one way "of mitigating the anthropogenic aberrations of the ecosphere" (Schellnhuber 1999: C23).

One year later, in 2000, Paul Crutzen and Eugene Stoermer suggested naming the current geological epoch the "Anthropocene." Due to the substantial influence of human activities on climate and other environmental processes, they asserted that "mankind will remain a major geological force for many millennia," necessitating a "world-wide accepted strategy leading to sustainability of ecosystems against human induced stresses" (Crutzen and Stoermer 2000: 18). While Crutzen and Stoermer made no mention of geoengineering, Crutzen reiterated their proposition in an article for *Nature* in 2002. In this essay, he explicitly related the idea of sustainable management in the Anthropocene to geoengineering, stating that a "daunting task lies ahead for scientists and engineers to guide society towards environmentally sustainable management during the era of the Anthropocene," and suggesting that this "may well involve internationally accepted, large-scale geoengineering projects, for instance to 'optimize' climate" (Crutzen 2002: 23).

From a green governmentality perspective, the question arises how the domain of human–nature relations is conceptualized and thereby rendered governable along particular logics or rationalities. In the systems perspective of earth system science, humans appear as a component of the earth system (a "geological force," as the notion of the Anthropocene implies) and as a collective global subject. This frame "positions researchers as metaphorical engineers whose job it is to help people cope with, or diminish, the Earth system perturbations unintentionally caused by their collective actions" (Castree et al. 2014: 764). Accordingly, John Lawton (2001: 1965) posits that "The greatest challenge for the new discipline [of earth system science] ... is to provide prescriptions that will reverse current human abuse of planet Earth, signposting routes to a sustainable future." In this thinking, there is little potential for conflicts about goals or about the appropriate means to achieve them. Natural *and* social matters become primarily scientific concerns, a move reflected also in Schellnhuber's vision of a "mathematical sustainable-development ethics" that would provide "a rigorous common formalism" to overcome the "lack of systematics in the overall sustainability debate" (Schellnhuber 1999: C23; see also Schellnhuber and Kropp 1998).

The problem with such framings is that they gloss over some of the most difficult challenges of contemporary environmental politics, such as struggles about power relations, violence and exclusion, political economy, conflicting priorities, and the status of scientific knowledge itself. In the language of optimization and rational management, politics is black-boxed and, because it fails to "function" according to scientific prescription, appears tedious and inefficient. The way that human–nature relations are conceptualized in earth system science is thus biased towards an expert-led and technical mode of governance, explaining the appeal of geoengineering; it holds the promise of control over global environmental parameters, presumably without the necessity to coordinate human activities (i.e., without having to open the black box of politics).

The idea of geoengineering is closely related to the development of an earth system science perspective on human–nature relations. This does not mean, however, that SRM appears as a simple solution to environmental problems in this view. Climate and earth system science have been at the forefront of examining and disentangling the manifold interactions and feedbacks in the climate system, emphasizing its inherent complexity. The question thus becomes how the complexity of the climate system might be rendered "calculable" and amenable to prediction and control. Here, computer simulation plays a key role as numerical models have become the central epistemic tool of earth system science. The next section shows that they provide a space in which ideas for SRM can be performed to produce quantitative images of climate futures. These supply the calculatory background for visions of technological climate control and optimization.

Virtual technology: solar radiation management in climate models

In 2010, Alan Robock and colleagues pointed out that a confined test of the climate consequences of solar geoengineering would be impossible without actually building and implementing the technologies on a global scale (Robock et al. 2010). Interventions designed to produce a detectable climate effect on a limited area would have repercussions beyond that area because the climate system is globally interconnected; and due to the large number of processes and the range of scales involved – from the seconds of cloud droplet formation to the centuries of carbon dioxide uptake in the oceans – the climate system cannot be rebuilt in a laboratory. As a consequence, computer simulations are central to the generation of knowledge about SRM.

The impossibility of studying the global climate system experimentally remained a major limitation to atmospheric science until the second half of the twentieth century and was overcome only with the emergence of new mathematical and computational tools that allowed for a numerical approximation of the theoretical equations used to describe global atmospheric circulation. Numerical climate models are translations of physical theories and assumptions into mathematical equations and ultimately computable code. Due to these translations, the complexity of the climate system, and the limited spatial and temporal resolution of climate models, they rely on numerous simplifications and approximations in order to create a system that behaves in a similar way to the planetary climate (Gramelsberger 2011). The high number of variables and processes included in today's models of the climate system means that a single scientist cannot overview and understand everything that happens in the simulation. The model itself thus remains "opaque" and in many aspects a black box, even to the modeller, who can take on the role of a user who provides the computer with a scenario and studies the outcome (Lenhard 2011). A scenario consists of a set of assumptions – for example, about the amount of GHG that will be emitted in the future or the application of SRM. Based on the scenario, the model generates a dataset that can be analysed and interpreted.

Climate simulations are internally consistent, but also uncertain because models are imperfect representations of their target system and because scenario design can go beyond realistic or plausible assumptions. Consequently, the boundaries between "science" and "fiction" become effectively blurred. Simulations are enacting a play, or a story; the model performs a set of assumptions provided by the scenario in the form of an "if ... then" narrative. As Isabell Stengers (2000: 136) writes:

> The art of simulating is that of the screenwriter: to put a disparate multiplicity of elements onstage; to define, in a mode which is that of a temporal, narrative "if ... then," the way these elements act together; and then to follow the stories that are able to engender this narrative matrix.

Against this background, simulation studies on SRM can be considered as producing narratives about a speculative technology; they indicate that *if* SRM was designed in a particular way, *then* the projected effects could be expected. Climate models thus provide a powerful narrative tool in the context of SRM.

They provide the means through which SRM technologies can be envisioned, allowing scientists to create a virtual analogue to a yet-to-be-built technological system; and by performing this "virtual technology," they produce datasets that allow for quantitative comparisons between different potential climate futures.

The first computer simulation study on the climate impacts of SRM appeared in 2000 (Govindasamy and Caldeira 2000) and since the beginning of the recent discussion in 2006 more than 50 studies have been published in peer-reviewed literature that consider the climate effects of different geoengineering methods based on climate model simulations. Ideas that have been examined include, among others, increasing the reflective properties of plants (Singarayer et al. 2009), the brightening of marine clouds (Jones et al. 2009; Rasch et al. 2009), particle injections into the stratosphere (Robock et al. 2008), and brightening large land surfaces, such as deserts (Irvine et al. 2011). It is not a precondition to the simulation whether any of this would be technically or physically viable; beyond a few thought experiments and back-of-the-envelope calculations, no SRM technology exists as yet. The modeller simply prescribes, for example, an increase in the optical depth of the stratosphere or an increased average concentration of cloud droplets. The easiest way of simulating SRM is to adjust the solar constant within the model, a single line of code that defines how much sunlight on average reaches the model earth. The virtual technology designed by the climate scientist is then included in a scenario that outlines how this virtual technology is applied.

The design of a scenario presupposes a formulation of a question or a problem. This formulation can be based on scientific considerations – for example, the need for a sufficient signal-to-noise ratio in order to distinguish between effects from SRM and other variables that cause climate variability – but often the scenario also entails a proposition, a hint towards the potential goals that could be pursued by SRM. Thus, for example, a majority of simulation studies have looked at the climate effects that would result from different SRM methods if they were used to offset carbon dioxide-induced global mean temperature rise (Kravitz et al. 2013), underlining the prevalent framing of geoengineering as a way to offset the effects of anthropogenic global warming. However, with the degree of control over sunlight provided by climate models, geoengineering opens a new range of imaginable technologies and climate states, and a series of ideas for other potential goals have surfaced based on simulation studies. MacMynowski (2009), for instance, examines whether it would be possible – in theory – to use SRM to control El Niño, while Latham et al. (2012) ask whether cloud brightening could be applied to weaken hurricanes. The role of scenario design in simulation studies is thus twofold. On the one hand, it allows scientists to learn about general features in the behaviour of the climate system. On the other, it appears that ideas of technological climate control do not precede their simulation; the iterative process of designing virtual technologies, performing scenarios, and studying the outcome produces storylines of global climate interventions. This constructive process is part of the practice of climate modelling, and while the representations of SRM in these models could be made more "realistic" if such technologies were to be developed, the converse is also true; simulation studies outline how SRM would have to be done in order to realize the goals prefigured in the scenarios.

Simulation studies thus create narratives about technological interventions in the climate system as a novel way of "governing" global environmental processes. From a governmentality perspective, the form that these narratives take is interesting. Simulation studies of SRM produce images of the future as datasets of climate variables that can be used for quantitative comparisons, making them accessible to particular ways of governing the problem. This is reflected, for example, in propositions to consider geoengineering as an "optimization" challenge:

> One approach to thinking about geoengineering is to ask "What kind of climate do we want?" and then ask "What pattern of radiative forcing from stratospheric aerosols would come closest to achieving that desired climate state?" This involves treating geoengineering as an optimization problem by defining "climate goals" and an "objective function" that measures how closely those goals are attained.
>
> *(Ban-Weiss and Caldeira 2010: 1)*

The datasets of environmental change provided by computer simulations thus provide the basis on which a quantitative assessment of different futures becomes viable. However, climate model-based projections for SRM hold information only about environmental variables. Several studies have thus begun to combine such projections with social indicators or economic models to extend the consideration to the social and political implications of SRM.

Virtual politics: calculating the social and political implications of SRM

The previous sections have outlined how SRM is related to ideas of global environmental management within the context of earth system science and how narratives of global climate control and optimization become possible by simulating "virtual technologies" in computer models. In the context of SRM, computer simulation studies are the primary tool by which different environmental futures are being generated, studied, and made calculable. But how is society accounted for? This section critically examines how simulation studies become the basis not only for calculating "nature," but also for assessments and claims about the social and political implications of SRM. Three interrelated aspects are considered: the construction of a "virtual ethics" in discussions about the distributional effects of SRM; the calculation of social "impacts" based on spatially and environmentally defined indicators; and the vision of centralized rational control.

A major concern within the discussion on SRM has focused on the heterogeneous effects on global temperature and precipitation patterns. The proposition that SRM would lead to an unequal distribution of impacts has in turn become a focal point for model-based assessments of regional disparities (Robock et al. 2008; Irvine et al. 2010). Several studies have since considered "optimization" of SRM as a way to minimize such regional disparities. The study by Ban-Weiss and Caldeira (2010), cited above, compares a low-carbon dioxide control simulation with a simulation of high-carbon dioxide conditions and an adjustable load of sulfate particles in the stratosphere. An optimal climate state is defined as a minimal aggregated deviation of climate indicators from the control simulation.[1] The study then modulates the spatial distribution of stratospheric aerosols based on a mathematical formula in order to approximate the optimization target as well as possible. In another study that considers SRM as an optimization problem, Moreno-Cruz et al. (2012: 659) combine climate model output data with an assessment model to provide a simple framework by which "[p]olicy- and decision-makers can compare different [SRM] proposals relative to the best case scenario for a given social objective." Going beyond a consideration of physical indicators of climate change, the authors include socio-economic indices in their consideration:

> To analyse differences in interregional preferences in a way that is impact-relevant, we weight changes in temperature and precipitation using data to represent three different social objectives: egalitarian, where each region is weighted by population (People); utilitarian, where each region is weighted by its economic output (US$ billion); and ecocentric, where each region is weighted in terms of Area (km^2).
>
> *(Moreno-Cruz et al. 2012: 650)*

Based on their results, the authors conclude that inequalities due to SRM "may not be as severe as it is often assumed" and that "a globally optimal level of SRM can compensate for a large proportion of damages at a regional level" (Moreno-Cruz et al. 2012: 661).

Similar attempts to assess regional inequalities based on climate model results have been made (Ban-Weiss and Caldeira 2010; Irvine et al. 2010; MacMartin et al. 2012) – for example, to answer the question whether SRM could provide an incentive for the formation of exclusive international coalitions (Ricke et al. 2013). They show that model outcomes serve not only to anticipate environmental changes caused by SRM, but also to evaluate its social implications. In assessments of regional inequalities, this has led to what could be termed a "virtual ethics." Since visions of the future are expressed primarily through computer simulation results, formulations of ethical concerns, such as distributional justice, need to be made compatible to these results in order to be considered in the calculation and to become part of the narrative. However, the terms and methods used by earth system science are often incompatible with research on social and political dimensions of human–nature interactions, partially because the latter are not as amenable to quantitative prediction. Climate, in turn, "becomes the one 'known' variable in an otherwise unknowable future," leading to an overemphasis on environmental change in visions of the future (Hulme 2011: 249).

As a result, while much effort is put into accounting for the complex ecosystem interactions of the earth system, humans tend to be reduced to static and spatially defined indices, such as population or economic output, in order to match the data structure of climate models and add a little "social flavour" to the discussion. The focus remains on projections of ecological variables used to determine social impacts, neglecting, for example, economic interdependencies, different cultural perspectives, political conflicts, or social networks that are at least as relevant to the ways in which people conceive of and cope with environmental change (see Barros et al. 2014). In other words, the centrality of computer simulations to SRM research easily leads to an ecological or "climate reductionism" in which model projections are used "as universal predictors of future social performance and human destiny," as Hulme (2011: 249) critically reflects. Furthermore, just like a currency, the use of single, globally applied indicators renders everything comparable, or, as Donna Haraway (1988: 580) writes, "what money does in the exchange orders of capitalism, reductionism does in the powerful mental orders of global sciences." In the eyes of the optimizing decision-maker, it makes no difference *where* a particular change occurs. As a consequence, and somewhat paradoxically, much of the discussion about regional disparities from SRM remains indifferent to the uneven geographies of climate change – that is, the uneven distributions of coping capacities, differentiated responsibilities, or unequal power relations.

Model-based optimization studies provide a rationale by which technological interventions in the climate system can be assessed as quantitative comparisons of climate futures. Implicit in such narratives of optimization is also a diagram of decision-making. As discussed above, the ideal of a global rational decision-maker has been part of an earth system science perspective on human–nature relations more generally. In simulation studies, this subject position reappears in the role of the modeller who carries out experiments with the climate and assesses the consequences of his or her actions. This position entails a deliberate break with, or distancing from, a human perspective. According to Schellnhuber (1999: C20), models can be viewed as "macroscopes," tools that are "giving Earth-system scientists an objective distance from their specimens" in order to "achieve 'holistic' perceptions of the planetary inventory, including human civilization." Computer simulations thus extend what Haraway (1988) has called the "god trick"; they allow scientists to free themselves of "cognitive comfort" (Schellnhuber 1999: C20) by pretending to see "everything from nowhere" (Haraway 1988: 581), and, in the case of climate modelling, prolong this vision into the future. The results are narratives about the potential for rational climate control that feature a single decision-maker who acts from the outside of sociopolitical struggles in the name of humanity.

The self-imposed distance of an earth systems perspective in geoengineering causes a widening gap between model-based visions of the future as a blueprint for expert-led climate control and alternative conceptions of environmental and technological change that place more emphasis on the role of human relations within these changes. As Noel Castree and colleagues (2014: 763) point out, much of global environmental change science continues to endorse "a stunted conception of 'human dimensions' at a time when the challenges posed by global environmental change are increasing in magnitude, scale and scope." But how could geoengineering become a topic of broader engagement without requesting that participants of such an engagement first accept the scientific framings that underlie the concept? It is to this question that I now turn.

Conclusions

Most academics working on geoengineering will be able to recall a situation where a brief mentioning of the concept yielded surprised looks of incomprehension and scepticism. Fertilizing the ocean? Sulphur in the stratosphere? Many, the author included, would soon find themselves explaining recent results from the scientific literature and the difficulty of weighing the risks of unmitigated global warming against the risks of SRM, teaching their counterpart about the state of research and stressing either that geoengineering involves large dangers (if the other readily endorses the promise of a technological "solution") or should not be dismissed out of hand (in the opposite case). This impulse to elucidate and clarify shows how deeply the notion of geoengineering is embedded into scientific modes of knowledge production and understanding human–nature relations. Although discussions tend to turn towards questions of ethics and politics over the course of the conversation, it shows how strongly geoengineering is presupposed on framing the problem of climate change as one of physical transformation and rational decision-making.

Research that does not readily follow that succession of explanation and that does not adopt the prevalent scientific framing as a natural starting point for engaging with geoengineering could help to broaden the discussion. The environmental social sciences and humanities have a lot to contribute to questions of how people understand, cope with, and influence environmental change. Putting more effort into understanding how the idea of technological interventions into the climate relate (or fail to relate) to the way that people think about and interact with the atmosphere is an obvious starting point for further interdisciplinary enquiry. Through such discussion, as Hulme (2011) suggests, society could be put back into the future. Rather than collapsing everything into a story of rational climate control, such an endeavour could and should "engender plural representations of Earth's present and future that are reflective of divergent human values and aspirations" (Castree et al. 2014: 763).

More diverse accounts of technology and politics in the Anthropocene could also "insure publics and decision-makers against overly narrow conceptions of what is possible and desirable as they consider the profound questions raised by global environmental change" (Castree et al. 2014: 763). Importantly, much research in the environmental humanities and social sciences does not aim for improved prediction and controllability of human–nature relations, but problematizes social relations that maintain, organize, and change vulnerabilities, access to resources, and regimes of knowledge production and decision-making. Researchers here may be particularly critical of the idea of geoengineering, but the reason for this may derive less from a lack of understanding of scientific "facts" and more from decade-long insights into the machinery of contemporary environmental change governance that almost everywhere claims to operate according to "rational" or efficient (market) mechanisms, but frequently discriminates against already marginalized communities, effectively maintaining existing power relations rather than challenging them. After all, the least plausible part of the scientific fiction of geoengineering may not be the viability of technological control over global environmental parameters, but the notion of a human agent employing such technology on the basis of sound scientific calculations and well-considered ethical principles.

Note

1 More precisely, the optimization task in the study is defined as a minimization of the root-mean-square deviation of regional temperature values or regional values for precipitation minus evaporation. The authors point out that the particular goals "were chosen for illustrative purposes only" (Ban-Weiss and Caldeira 2010: 3).

References

Ban-Weiss, G.A. and Caldeira, K. (2010) "Geoengineering as an optimization problem," *Environmental Research Letters*, 5(3), http://iopscience.iop.org/1748-9326/5/3/034009/pdf/1748-9326_5_3_034009.pdf, accessed 22 March 2015.

Barrett, S. (2008) "The incredible economics of geoengineering," *Environmental and Resource Economics*, 39: 45–54.

Barros, V.R., Field, C.B., Dokken, D.J., Mastrandrea, M.D., Mach, K.J., Bilir, T.E., Chatterjee, M., Ebi, K.L., Estrada, Y.O., Genova, R.C., Girma, B., Kissel, E.S., Levy, A.N., MacCracken, S., Mastrandrea, P.R., and White, L.L. (eds) (2014) *Climate Change 2014: Impacts, Adaptation, and Vulnerability, Part B: Regional Aspects: Contribution of Working Group II to the Fifth Assessment Report of the Intergovernmental Panel on Climate Change*, Cambridge and New York: Cambridge University Press.

Braun, B. and Wainwright, J. (2001) "Nature, poststructuralism, and politics," in N. Castree and B. Braun (eds) *Social Nature: Theory, Practice, and Politics*, Oxford: Blackwell, pp. 41–63.

Castree, N., Adams, W.M., Barry, J., Brockington, D., Büscher, B., Corbera, E., Demeritt, D., Duffy, R., Felt, U., Neves, K., Newell, P., Pellizzoni, L., Rigby, K., Robbins, P., Robin, L., Rose, D.B., Ross, A., Schlosberg, D., Sörlin, S., West, P., Whitehead, M.,

and Wynne, B. (2014) "Changing the intellectual climate," *Nature Climate Change*, 4: 763–8.
Cosgrove, D. (2006) *Geographical Imagination and the Authority of Images*, Stuttgart: Franz Steiner Verlag.
Crutzen, P.J. (2002) "Geology of mankind," *Nature*, 415: 23.
Crutzen, P.J. (2006) "Albedo enhancement by stratospheric sulfur injections: A contribution to resolve a policy dilemma?," *Climatic Change*, 77: 211–20.
Crutzen, P.J. and Birks, J.W. (1982) "The atmosphere after a nuclear war: Twilight at noon," *Ambio*, 11: 114–25.
Crutzen, P.J. and Stoermer, E.F. (2000) "The 'Anthropocene,'" *IGBP Newsletter*, 41: 17–18.
Edwards, P.N. (2011) "History of climate modeling," *Wiley Interdisciplinary Reviews: Climate Change*, 2: 128–39.
Escobar, A. (1996) "Construction nature: Elements for a post-structuralist political ecology," *Futures*, 28: 325–43.
Fleming, J.R. (2006) "The pathological history of weather and climate modification: Three cycles of promise and hype," *Historical Studies in the Physical and Biological Sciences*, 37: 3–25.
Foucault, M. (2007) *Security, Territory, Population*, Basingstoke: Palgrave Macmillan.
Govindasamy, B. and Caldeira, K. (2000) "Geoengineering earth's radiation balance to mitigate CO_2-induced climate change," *Geophysical Research Letters*, 27: 2141–4.
Gramelsberger, G. (2011) "What do numerical (climate) models really represent?," *Studies in History and Philosophy of Science Part A*, 42: 296–302.
Haraway, D. (1988) "Situated knowledges: The science question in feminism and the privilege of partial perspective," *Feminist Studies*, 14: 575–99.
Hulme, M. (2011) "Reducing the future to climate: A story of climate determinism and reductionism," *Osiris*, 26: 245–66.
Intergovernmental Panel on Climate Change (IPCC) (2013) *Climate Change 2013: The Physical Science Basis: Contribution of Working Group I to the Fifth Assessment Report of the Intergovernmental Panel on Climate Change*, Cambridge and New York: Cambridge University Press.
Irvine, P.J., Ridgwell, A., and Lunt, D.J. (2010) "Assessing the regional disparities in geoengineering impacts," *Geophysical Research Letters*, 37, http://onlinelibrary.wiley.com/doi/10.1029/2010GL044447/full, accessed 22 March 2015.
Irvine, P.J., Ridgwell, A., and Lunt, D.J. (2011) "Climatic effects of surface albedo geoengineering," *Journal of Geophysical Research: Atmospheres*, 116, http://onlinelibrary.wiley.com/doi/10.1029/2011JD016281/full, accessed 22 March 2015.
Jones, A., Haywood, J., and Boucher, O. (2009) "Climate impacts of geoengineering marine stratocumulus clouds," *Journal of Geophysical Research: Atmospheres*, 114, http://onlinelibrary.wiley.com/doi/10.1029/2008JD011450/full, accessed 22 March 2015.
Kravitz, B., Robock, A., Forster, P.M., Haywood, J.M., Lawrence, M.G., and Schmidt, H. (2013) "An overview of the Geoengineering Model Intercomparison Project (GeoMIP)," *Journal of Geophysical Research: Atmospheres*, 118, http://onlinelibrary.wiley.com/wol1/doi/10.1002/2013JD020569/full, accessed 22 March 2015.
Latham, J., Parkes, B., Gadian, A., and Salter, S. (2012) "Weakening of hurricanes via marine cloud brightening (MCB)," *Atmospheric Science Letters*, 13: 231–7.
Lawton, J. (2001) "Earth system science," *Science*, 292: 1965.
Lenhard, J. (2011) "Artificial, false, and performing well," in G. Gramelsberger (ed.) *From Science to Computational Sciences*, Zürich: Diaphanes, pp. 165–76.
Lovbrand, E., Stripple, J., and Wiman, B. (2009) "Earth system governmentality," *Global Environmental Change*, 19: 7–13.

MacMartin, D.G., Keith, D.W., Kravitz, B., and Caldeira, K. (2012) "Management of trade-offs in geoengineering through optimal choice of non-uniform radiative forcing," *Nature Climate Change*, 3: 365–8.

MacMynowski, D.G. (2009) "Can we control El Niño?," *Environmental Research Letters*, 4, http://iopscience.iop.org/1748-9326/4/4/045111/pdf/1748-9326_4_4_045111.pdf, accessed 22 March 2015.

Marchetti, C. (1976) "On geoengineering and the CO_2 problem," IIASA Research Memorandum RM-76-017, Laxenburg: International Institute for Applied Systems Analysis.

Miller, C. (2004) "Climate science and the making of a global political order," in S. Jasanoff (ed.) *The Co-Production of Science and Social Order*, New York: Routledge, pp. 46–66.

Moreno-Cruz, J.B., Ricke, K.L., and Keith, D.W. (2012) "A simple model to account for regional inequalities in the effectiveness of solar radiation management," *Climatic Change*, 110: 649–68.

Newton, P. (1999) "A manual for planetary management," *Nature*, 400: 399.

Perry, A.H. (1985) "The nuclear winter controversy," *Progress in Physical Geography*, 9: 76–81.

Preston, C.J. (2013) "Ethics and geoengineering: Reviewing the moral issues raised by solar radiation management and carbon dioxide removal," *Wiley Interdisciplinary Reviews: Climate Change*, 4: 23–37.

Rasch, P.J., Latham, J., and Chen, C.C. (2009) "Geoengineering by cloud seeding: Influence on sea ice and climate system," *Environmental Research Letters*, 4, http://iopscience.iop.org/1748-9326/4/4/045112/pdf/1748-9326_4_4_045112.pdf, accessed 22 March 2015.

Ricke, K.L., Moreno-Cruz, J.B., and Caldeira, K. (2013) "Strategic incentives for climate geoengineering coalitions to exclude broad participation," *Environmental Research Letters*, 8, http://iopscience.iop.org/1748-9326/8/1/014021/pdf/1748-9326_8_1_014021.pdf, accessed 22 March 2015.

Robock, A., Oman, L., and Stenchikov, G.L. (2008) "Regional climate responses to geoengineering with tropical and arctic SO_2 injections," *Journal of Geophysical Research: Atmospheres*, 113, http://onlinelibrary.wiley.com/doi/10.1029/2008JD010050/full, accessed 22 March 2015.

Robock, A., Bunzl, M., Kravitz, B., and Stenchikov, G.L. (2010) "A test for geoengineering?," *Science*, 327: 530–1.

Rose, N. and Miller, P. (1992) "Political power beyond the state: Problematics of government," *British Journal of Sociology*, 43: 173–205.

Royal Society (2009) *Geoengineering the Climate: Science, Governance and Uncertainty*, London: The Royal Society.

Rutherford, S. (2007) "Green governmentality: Insights and opportunities in the study of nature's rule," *Progress in Human Geography*, 31: 291–307.

Schellnhuber, H.J. (1999) "'Earth system' analysis and the second Copernican revolution," *Nature*, 402: C19–C23.

Schellnhuber, H.J. and Kropp, J. (1998) "Geocybernetics: Controlling a complex dynamical system under uncertainty," *Naturwissenschaften*, 85: 411–25.

Singarayer, J.S., Ridgwell, A., and Irvine, P. (2009) "Assessing the benefits of crop albedo bio-geoengineering," *Environmental Research Letters*, 4, http://iopscience.iop.org/1748-9326/4/4/045110/pdf/1748-9326_4_4_045110.pdf, accessed 22 March 2015.

Stengers, I. (2000) *The Invention of Modern Science*, Minneapolis, MN: University of Minnesota Press.

Victor, D.G., Morgan, M.G., Apt, J., Steinbruner, J., and Ricke, K. (2009) "The geoengineering option: A last resort against global warming?," *Foreign Affairs*, 88: 64–76.

von Storch, H. (2004) "A global problem," *Nature*, 429: 244–5.

Weart, S. (2010) "The development of general circulation models of climate," *Studies in History and Philosophy of Science Part B: Studies in History and Philosophy of Modern Physics*, 41: 208–17.

9
BIOFUELS
Climate solution or environmental pariah?

James Smith and Shaun Ruysenaar

Introduction

In 2009, John Beddington, the UK's Government Chief Scientific Advisor at the time, coined the phrase "perfect storm" to describe an impending global confluence of food, water, and energy insecurity, driven by climate change and unsustainable growth (Beddington 2009). Prior to this announcement, biofuels had risen to prominence as a panacea for a range of ills, many directly relating to this coming storm. Their promotion and currency were fuelled by "win–win" narratives; biofuels were seemingly more sustainable and would supplement rather than substitute for fossil fuels. They would not alter the existing socio-technical assemblage that surrounds fossil fuels (including the motor vehicles most Westerners use every day). Attached to this central theme came the potential for rural development in less developed countries (with a new market for their agricultural production), greenhouse gas (GHG) neutrality, national fuel security, and a range of claims that stoked enthusiasm.

However, the euphoria and seemingly collective enthusiasm quickly turned to an outcry. Questions of water availability, negative energy balances, loss of arable land, and subsequent loss of food production to fuel production came to the fore to reframe a potential solution as a series of environmental and developmental problems. In the space of a few years, biofuels have evolved from a technological niche product, to a possible solution, to an array of global ills, to a "crime against humanity," according to the UN Special Rapporteur on the Right to Food (quoted in Lederer 2007).

This chapter focuses on the recent development of biofuels as both a solution to and perhaps also a driver of climate change (as a component of policy "perfect storms"). Reframing biofuels – or, perhaps more precisely, shifting from the macro-narratives suggested by the perfect storm storyline to tracing multiple causal chains and mapping their implications – may allow us to generate governance systems that are themselves framed by more nuanced conceptualizations of risk and reward, and the trade-offs therein. We draw on the example of South Africa to highlight the need for stronger connections between systems of governance and the knowledge economy. These will be necessary if effective and sustainable decisions are to be made about biofuel development, and if they are truly to be part of a trajectory of development that acknowledges and manages climate change.

Biofuel basics

"Biofuels" are liquid fuels that are directly derived from renewable biological resources, especially from purpose-grown energy crops (the term "agro-fuels" has been used to single out the large-scale, commercially produced biofuels). Biofuels, and bioenergy more generally, are nothing new to developing countries. Two and a half billion of the world's poorest people rely on bioenergy every day, and biofuel production has been practised for some decades in Africa, especially in Mali, where jatropha (a shrub that produces seeds that can be used in the production of biodiesel) has been widely grown (ENDA 2008). What is new is that, largely as a result of climate change policies in the European Union and surplus corn supply in the United States, there have been recent massive increases in commercial biofuels or agro-fuels grown predominantly for use in the developed countries. Between 2002 and 2006, the amount of land used to grow biofuel crops quadrupled and production tripled (Coyle 2007). In 2011, over 60 per cent of all land acquired in Africa was found to be for biofuel production (Schoneveld 2011). Virtually all of the commercially available agro-fuels are "first-generation" biofuels that are produced from starch- or sugar-rich crops, such as maize or sugarcane (for bioethanol), or oilseeds, such as rapeseed, soy, palm, or jatropha (for biodiesel). Many of these crops are edible, which, in part, given fears that food is being diverted into fuel production, has prompted research into non-edible biofuels that (potentially at least) appear to pose less of a threat to the production of food crops. For the moment, however, investment is in edible crops as the sources for biofuels.

The globally integrated biofuels network and its investment-driven transition from regional assemblages of, for example, cooperatives and small farmers to, increasingly, transnational companies and national governments highlight the inequity between where energy is consumed and where energy – via biofuels – may originate. Simple calculations indicate that all of the vegetation in the United States contains only one-third of the energy consumed in one year in that country (Moore 2008). Furthermore, these figures do not account for any energy expended in the production, processing, or transportation of biofuels (Pimentel 2004, cited in Moore 2008). These statistics highlight two issues: first, the limits of the potential of first-generation biofuels to address the world's energy demands; and second, a pattern of consumption that would inevitably mean that these demands would have to be met elsewhere, especially in the global tropics, where land and labour are relatively cheap and the potential for biomass production is greatest. This geography both represents the potential for economic development and has implications for agricultural production in regions of the world where agriculture is already under stress (and likely to become more stressed, due to climate change).

Nevertheless, investment in biofuels continues. Energy security is a consideration; oil is running out certainly, but countries want to secure their access to alternatives before the wells run dry, especially given the political and economic instability in several regions of the world where oil is produced. Global rural development is given as another reason for continued investment in biofuels. Alongside this, the production of biofuels and diversification into alternative energy sources more broadly are seen as strategic opportunities for rural areas – in developed and developing countries alike – to grow their economies, diversify their income sources, and create jobs. In OECD countries, the overproduction of agricultural commodities, low prices, underutilized land, low farm incomes, and powerful farm lobbies have created optimum conditions in which new agricultural commodity markets can be cultured. In the United States and the European Union, existing heavy agricultural subsidies can be readily recalibrated towards the growing of crops for biofuel production. In developing countries, too, there are economic and developmental drivers of biofuel production, primarily those of reduced oil imports, rural development opportunities, and subsequent opportunities for exports and income. Additionally, a lack of access to energy is increasingly framed as a fundamental impediment to development and consequently as a cornerstone of developing countries' poverty alleviation strategies (FAO 2000).

Biofuels represent a global techno-optimism and they are a component of growing systemic interconnectedness, and all that that entails. They exist within a policy world where science, properly understood and governed, will help us to avoid difficult resource decisions, development challenges, or environmental implications. To an extent, this is simply the politics of issue shifting, or generation displacing – shuffling the problem elsewhere through space and time as we ponder new technological interventions that might provide cleaner, cheaper energy. In developing countries there is a parallel process, with technological innovation being increasingly touted as a driver of poverty alleviation. There is an overriding narrative that we must improve the harnessing of science and technology to these ends – environmental and developmental – and that in so doing we will perhaps sidestep extraordinarily complex problems with beguilingly neat solutions (see also Wiertz's discussion, Chapter 8, this volume).

One of the most striking features of biofuels as a global solution to the intersecting problems of a lack of access to energy and the risks of climate change and uneven development is their potential entirely to reshape livelihoods, patterns of resource consumption, environments, and agro-food production systems. Yet, there is a cost to every benefit, and uncertainty is often hidden in relation to the technological promise. The challenge is that current governance solutions, like those of climate change, express the existing political economies and development rationales that created the perfect storm in the first place (see also Glover's analysis, Chapter 2, this volume). Mass displacement of people from their land in order to grow crops for foreign fuel, food, and feed markets appears at odds with current developmental imperatives and seems stuck in paradigms reflecting old-school geopolitical resource control and primary commodity export-driven economic strategies. This may well be the flipside of the global techno-optimism framing of biofuels.

In some respects biofuels are simple technologies. We are simply deriving energy from plants, through seed oil or biomass, primarily to put fuel in car engines. Future, better technologies may unlock efficiencies or new ways of deriving energy from plants, but for now the basic principle remains. In other respects biofuels are extremely complex. They are being developed within complex political, economic, and environmental systems and promoted by an increasingly fragile globally integrated food–feed–fuel agricultural commodity market in which their production in itself generates further complexity. Biofuel production generates couplings between agricultural systems, international markets, petrochemical companies, consumers, and producers. Whereas fuel used to be a critical input to agricultural production and distribution, this apparently neat system is being made far more complicated by fuel now being an input and a product. These couplings have implications of their own, in terms of both immediate impacts and our ability to foresee future impacts, both positive and negative (Perrow 1999).

Biofuels have been almost simultaneously articulated as a means to secure cleaner energy and as a pathway towards economic development, on the one hand, and an environmentally devastating proposition that will further entrench inequalities, on the other. These widely differing viewpoints highlight two important issues around biofuels as an effective response to climate change. The oscillation between solution and driver is fundamentally important, and this chapter argues that this oscillation has two connected dimensions. First, our scientific and technical knowledge regarding the impacts and emissions of biofuels is insufficiently developed. Second, the global (and often national) governance systems necessary to make effective choices around the development of biofuels are not effective in balancing risks and benefits at multiple levels.

The limits of technical knowledge: GHG "neutrality"

As we have noted above, there are seemingly many reasons to invest in biofuels. Much of the attraction lies in their perceived GHG neutrality. Not only can biofuels replace petroleum-based fuels, they can also help to mitigate climate change. As crops grow, they fix carbon from the atmosphere. When they are combusted as biofuels in engines, this carbon is simply released back into the atmosphere; hence, the net impact on atmospheric carbon is, in theory at least, neutral. In reality, however, biofuels are not GHG neutral. There is a range of emissions associated with all stages of their life cycle. Growing crops intensively, using nitrogen-based fertilizers and petroleum-powered farm machinery, processing, and transportation all require large inputs of energy, and that energy is usually provided by fossil fuels.

Of course, biofuels simply have to emit less GHG than fossil fuels for them to be beneficial in terms of tackling climate change, but it is far from clear that many of them do substantially cut net emissions in reality. Biofuels might appear intuitively sensible in many respects, but the reality is more complex, and we need analytical tools that allow us to make sensible, sustainable decisions about their use. We noted earlier the mismatch between the theoretical energy potential of US vegetation and US energy use. For example, a more realistic analysis would consider the energy sequestered within arable biomass in the United States as the upper limit, or offset the energy inputs required to farm and turn crops into fuel against fuel produced. For this "boundary setting" exercise to be meaningful, we would have to set tighter boundaries and, each time we did so, we would assert a value judgement, add complexity, and include risk assumptions within our analysis.

Currently, the most scientifically acceptable means to measure environmental impacts from "field to wheel" of biofuels is through life-cycle analyses (LCAs). LCAs are usually studied comparatively, in order to analyse which alternative energy source has the lesser environmental impact. More often than not, these analyses focus on the impacts that the production and use of biofuels might have on emissions of GHGs relative to emissions from the use of conventional petroleum-based transportation fuels (van der Voet et al. 2010).

A key concern in constructing LCAs is the realistic (and comparative) drawing of boundaries around the various inputs, outputs, and processes that make up the theoretical "life cycle" of biofuel production and use. Thus, an LCA cannot be considered a neutral analysis of the environmental impacts of various types of biofuels; rather, it represents a best guess of as complete a system as possible. There is a risk that neutrality may be forgone if one is intent on framing biofuels in a certain way, or making a particular point. The most striking feature of LCAs is the huge diversity of their results. A significant conclusion from a range of meta-analyses (see, for example, Whittaker et al. 2010) is that there are frequently wide ranges of results reported for a given biofuel and originating biomass source. Widely diverging results can be partially explained by different biomass sources, different agro-ecological contexts, different processing techniques, and other technical and contextual factors. However, different methodologies, dependent on different assumptions that underpin them, different (and new) datasets, and advancing technologies, also serve to skew analysis.

An important issue is that the majority of LCAs have been undertaken in a European or North American context. Given the emphasis on biofuel production in developing countries, studies in Europe or North America can provide only indicative or proxy results. The context-specific variability and breadth of input values in LCAs – for example, there is far more solar radiation in the tropics than in European agricultural areas – means country- or region-specific studies are far more valuable in determining meaningful results and making informed decisions. Of course, there will be always be limits to scientific models as reality is always more complex, but models must balance simplicity and robustness if they are to make meaningful contributions to policy- and decision-making. As an example of limited data availability for the decision-making process, *An Investigation into the Feasibility of Establishing a Biofuels Industry in the Republic of South Africa* (Biofuels Task Team 2006), which was described as a comprehensive study undertaken to inform the direction that the national biofuels policy would take, utilizes the example of rapeseed in the UK because no such analyses had been undertaken in the South African context. If any equivalent analyses did exist at the time, they were not included as part of the feasibility study. More detailed LCAs for the South African context were completed only a few years later, after many of the policy decisions had already been made. We will return to the use of such evidence in policy later.

Perhaps the most significant omission from many LCAs is the implication of both direct and indirect land use change. Vegetation plays a major role as a carbon sink, sequestering a fifth of human-made emissions every year. (When carbon is stored in plants and soils, it is called "biological carbon sequestration." Since this sequestration takes carbon dioxide out of the atmosphere, it is also known as a GHG "sink.") As new land is put into production – for example, through cutting down and burning trees, ploughing soil, and changing the type of vegetation – significant GHG emissions can result from releasing stored carbon. Typically, in comparing fossil fuels and biofuels, LCAs work on the assumption that growing biofuel feedstock sequesters carbon dioxide from the atmosphere. This sequestered carbon acts as a sufficiently large credit that overall GHG emissions from biofuels are lower than those from fossil fuels, even if the process of producing biofuels may in itself generate more emissions than mining fossil fuels. This "credit" needs to be understood in the context of what the land had been used for previously. Many studies work from the assumption that land was not previously used for growing crops or vegetation, which risks bias in effectively including carbon credits but ignoring carbon costs (Searchinger et al. 2009). In reality, farmers may plough grassland or slash and burn forests in order to plant crops for biofuels, releasing the majority of previously sequestered carbon back into the atmosphere. They may also choose to divert existing crops into biofuel production, which has similar implications as cropland may be expanded elsewhere to compensate for reduced food crop production.

LCAs have tended to operate on the assumption that any land on which biofuel feedstocks were grown was previously "set aside," thereby negating land use change emissions. This is clearly problematic. Destroying peatland may lead to emissions of several hundred tonnes of carbon per hectare. Emissions from the destruction of peatland in Southeast Asia have been estimated at about two billion tonnes of carbon per year. Demand for palm oil, for example, is the largest driver of peatland destruction in Indonesia (Royal Society 2008). Comparing the amount of emissions caused by clearing land with the emissions savings offered by biofuel crops offers cautionary tales in terms of how much net benefit there is to reducing climate change. Clearing Indonesian peatland and replanting it with palm oil would generate a carbon debt that would take 420 years to pay back (Fargione et al. 2008). These more complex, context-specific analyses appear to have little bearing on policy-making. For reasons we will discuss in the next section, this lack of connection between detailed analyses and policy may not be as surprising as one might think. Yet the implications are profound, given that decisions that lead to replacing peatland or tropical forest with biofuel crops are effectively irrevocable.

Prospective, ex ante studies by definition deal in uncertainties and assumptions (see, for example, Smith 2007). This also explains the wide range of results from LCA studies. Methodological and analytical differences play an important part as well. LCAs are almost by definition narrowly focused on tracking cumulative emissions through fuel-type value chains in order to understand whether one-for-one substitution of one fuel type for another makes sense. They typically do not analyse the consequences of policy actions or price changes. For example, if a biofuel could be produced more cheaply than a petroleum equivalent, that price differential might lead to greater fuel consumption, offsetting any environmental gains. To incorporate all possible feedbacks and interactions would be impossibly complex, yet carbon neutrality has become a significant rallying call in the overarching biofuels discourse. Whereas there are obvious limitations in our technical abilities to calculate such neutrality, what deserves equal discussion is how we govern technologies to mitigate any negative impacts and how decisions are made in situations where such neutrality is questionable. It is here, in the realms of governance, where the complexity of biofuels and their backward and forward linkages to other systems come to the fore.

The limits to governance: grappling with complexity

Our second argument suggests that current systems of government, particularly their approaches to governing complex technologies, are ill-equipped to deal with the implications of biofuels, let alone frame policy questions in a manner allowing for good governance in the first place. That is, if indeed we are to manage biofuels as a mechanism to mitigate climate change, do policy-makers ask the right questions and are they likely to receive the right answers? (See O'Lear, Chapter 7, this volume, for further discussion.) The situation becomes even more complicated given that policy-makers are facing multiple and yet relatively unclear pending crises, with climate change but one of them. Perhaps the most vexing issue that LCAs and other potentially useful analyses need to accommodate is that many value chains have inputs, outputs, and outcomes that cross multiple levels (for example, local versus global, short term versus long term) of different scales (for example, space and time).

When it comes to global issues such as climate change or integrated biofuels networks, the significance of multi-level perspectives should not be forgotten. Arthur Mol (2010: 61) argues that "conventional state environmental authorities have limited power and legitimacy to effectively regulate the sustainability of current global biofuels." He also points to the emergence of new governance frameworks, including private-sector "authorities" that should be neither condemned nor revered but demand greater scrutiny. One of the challenges here is how these new types of authority (and, more importantly, the old ones) are able to manage the confluence of complex systems that are encountered at the nexus of changing climates, complicated (and collapsing) food systems, and energy systems, and how the production of biofuels fits into this situation.

The concept of governance is itself wide-ranging, sometimes used to describe particular ways of governing and at others used as a collective term for all forms of governing (see Stoker 1998; Colebatch 2009). The inclusion of the private sector, as mentioned above, is evidence of increasingly network-based or "heterarchical" governing structures (Lyall et al. 2009). In simple terms, it implies that the act of governing no longer resides (or never did reside) wholly with government but increasingly involves a range of other actors; a continuum exists between government and governance. At one end of the continuum, government plays an overriding role. At the other, multiple stakeholders are involved (see Figure 9.1). Ultimately, governments continue to play a central role throughout; but it is worth noting some of the limitations to the global governance of biofuels, beyond the boundaries of the state (which exemplify the emphasis of the network governance approach).

One response to Mol's (2010) analysis, for example, is to suggest that global governance requires far more stringent and more consistently applied controls. The irrevocable nature of many of the implications of biofuel production – whether around land use or investment – means that control must come first. Trying to assert power to regulate sustainability in retrospect will not work if forests have already been cut down or technological lock-in has occurred. This is especially problematic in developing country contexts, where state power is

Governance "hierarchy" ↑ Increasing degree of government coercion ↓ Government "hierarchy"	Approaches	Tools
	Sermons, influence	Public campaigns and voluntary standards
	Rewards, "carrots"	Public campaigns, voluntary standards, and certification
	Punishment, "sticks"	Regulations and penalties

FIGURE 9.1 The governance–government continuum and instruments of governance.
Source: adapted from Lyall et al. (2009)

limited and the influence of private-sector investment powerful. However, the international governance of biofuels remains at the "voluntary" or persuasive end of the governance spectrum, generally relying on trust and persuasion along the standards and certification value chain, which includes both government and non-government actors. It is particularly difficult to ensure the desired results when using this approach (Lyall et al. 2009). Whereas many of these standards are state-based or governed by industry organizations (international and local) in "partnership" with state administrations, their coverage is quite limited and significant loopholes exist, especially in knowledge (see Fortin 2011). While more powerful states are able to wield some control as they have the authoritative capacity to do so, others — especially less developed countries — remain ill-equipped. The inability of these states to punish, or even persuade, means that comprehensive lists of sustainability criteria are merely wish-lists of future desires rather than codes of conduct for current practices.

In general, the lack of consensus over the criteria that define various standards further marginalizes their utility. Ultimately, while standards may maintain performance around components of biofuel production, they do not (and possibly cannot) govern the backward and forward linkages of the biofuels sector. They provide merely snapshot governance. The governance of global biofuels production and trade is thus in dire need of review and revision. First, there is a need to recognize the biased and skewed balance of power in the global political economy and build appropriate institutions to set and enforce rules and controls (Koning and Mol 2009). Second, it is important to reframe the discussion in terms of different scales and the interactions between and across them. Often, the focus is on institutions — the rules of the game. There are, however, jurisdictional, spatial, temporal, and other scales that are equally important but often ignored, mismatched, or too narrowly conceptualized (see Cash et al. 2006). The task is not an easy one. The strength of multilateral governance tends to be only as strong as its weakest state (or sometimes as weak as its strongest states; see Levidow 2013) and the process of policy-making itself presents further complications.

"Governing biofuels": policy in South Africa

We have noted that internationally institutional (and other) mechanisms remain insufficient to control the rapid uptake and risks associated with biofuels. In their absence, the only option is to rely on state-based approaches to protect (or neglect) national social, environmental, and economic interests. In this sense, governing biofuels is manifested through the articulation and implementation of public policy. As we also noted earlier, developing the analytical tools to generate the evidence that policy-makers require to make decisions is very difficult, with LCAs and their challenges being one example. This is not the only issue, however. There remains an overly instrumental perspective that assumes that policy-makers are both willing and able to use whatever knowledge exists, when often decisions are usually made in the face of incomplete and competing evidence and under intense time constraints, among other pressures. This scenario suggests we need to understand the importance of how policy questions are framed and answered (and, in the case of biofuels, how policies are often about technical solutions looking for policy problems).

In so doing we need to reframe how policy-making processes around biofuels and climate change are understood. Too often, policies are assessed for *what* they say (usually in comparison to recommendations of what people think policies should say), not for *who* had a say, and how certain framings of issues may supersede and marginalize others according to political or economic rather than social, environmental, or technical prerogatives. The latter point suggests that tracing and understanding the development of biofuels policies may provide greater insight into biofuel governance, with a greater need for analyses of how such policies are actually produced.

In looking at the South African context – that is, the development of the country's biofuels policies and the emerging biofuels regime – some interesting lessons come to the fore. To crystallize these lessons, we outline two main areas that highlight some of the tangible limits to governance in terms of, first, how policies are made and, second, how those policies relate to the practice of governance.

Given that policies and legislation are two of the fundamental features of government, it would appear that if we do not get policy "right," governance fails (or we begin to rely too much on other governance structures to fill the gaps). Much of the biofuels literature focuses on producing sufficient evidence to ensure the correct decisions are made, which results in well-defined instrumental policy. In instances where problems are of a strictly technical nature, this approach may work. It denies, however, the reality of most situations, in which policy-making is a political activity and problems are rarely only technical in nature. That is, even with biofuels being a technical solution of sorts, technocratic decision-making is inherently bound up in politics (Fischer 1990) and the way in which policy decisions are made in the real world is not so neatly defined. Problems and their solutions are socially constructed; they are not a matter of plugging in the data and retrieving the "correct" answer. For example, an overarching construction of small-scale farmers in South Africa – generally peasant farmers who were structurally marginalized during apartheid – is that they are disconnected from the mainstream or "first economy" and reside in a "second economy." This notion of two economies was entrenched at the highest political level after a speech in 2003 by former President Thabo Mbeki drew heavily on the concept of a dual economy (Philip 2010). The policy solution then becomes building bridges between the two so-called economies, when the reality is actually of one – extremely unequal – economy linked together in problematic ways. Nevertheless, following the hype surrounding their potential for agricultural development, biofuels became one such "bridge" between constructed economic realities.

In line with the general thrust of this book, we underline some of the limitations of current views of evidence-based policy (the underlying principle of getting the policy right) and contrast these with an alternative view of the politics of evidence in policy-making. In no way do we seek to undermine the importance of evidence in policy-making, but rather suggest that it is not the only element involved in policy-making and governance, and it would be naive to think that it is. Crucially, this stretches far beyond the debate over biofuels; the politics of evidence is at the heart of some of the confusion around climate change and the responses that are being taken towards it. For that reason, we suggest reframing the unequivocal, often unproblematized, promotion of evidence-based policy – the overarching normative framework through which policy-making is currently evaluated.

In questioning the idea of evidence-based policy-making, two practices may be dissected. The first is obtaining evidence in the first place. As noted above, in the face of the complexity surrounding climate change and biofuels, obtaining sufficient evidence is already problematic, given the number variables to be measured. These variables are sometimes hidden, they influence one another in complex ways, and there are unclear relationships between them (and those of other systems). Returning to our discussion of GHG neutrality, if environmental savings through reduced GHG emissions may be significant, it might be worth investing in research or incentives to encourage biofuel production. On the other hand, if biofuel production is of only marginal benefit, it may well be that the opportunity costs of investment do not add up, and we should focus our attention elsewhere. Developing the analytical tools to provide the evidence that policy-makers require is very difficult. There are limitations and trade-offs between accuracy and the extent of available data in analysing the true environmental costs of biofuel production. Scientifically, we may simply not have the knowledge or the data that we need fully to realize a workable analysis. Investment in biofuel production already appears to be outpacing high-quality research. Von Braun (2007) argues that even though many of the plans to expand biofuel production in developed and developing countries have been made with very little analytical basis, they have now become policy.

The second activity to dissect, given the potential for an information vacuum, is that interests, politics, and investments may exert excessive influence in shaping the context in which analyses are undertaken and interpreted. That is, room exists for both evidence-based policy-making and policy-based evidence-gathering; policy is not a linear process of gathering information and deciding on a solution, but an iterative and chaotic process where evidence can be manipulated, reinterpreted, or entirely neglected.

The aforementioned feasibility study in South Africa, undertaken by government officials and industry consultants to assess the potential for a biofuels industry in the country, illustrates a few of the reasons why evidence and the way in which it is framed should be questioned. The most obvious starting point is the supposed "independence" and objectivity of the scientist or expert, and of any evidence used in what might be called the knowledge–policy interface. While, on the one hand, there is the optimistic view that knowledge can speak truth to power (Pielke 2007; see also Yearley 2005; O'Lear, Chapter 7, this volume), it is equally possible that scientists and experts are not impartial or simply do not have the knowledge base to understand what impartiality is in a particular context.

In the case of the feasibility study, two points should be made. The first is that early critiques of the feasibility criteria (see Austin 2006) suggested some serious shortcomings in the type of data that could be collected in the specified timeframes and the possible exclusion of more detailed and relevant analyses of climate change, food security, and the real potential of the biofuels value chain to achieve the claimed rural development benefits. Second, there were perceived conflicts of interest as the involvement of consultants drawn largely from the petrochemical industry meant potentially significant biases informed the direction of the study itself. Indeed, the focus of discussions drifted from broad developmental approaches and niche products towards merging biofuel production with existing infrastructure.

Beyond the bias of the experts or scientists, Collingridge and Reeve (1986) suggest that the features of science-for-policy make it unworthy as an aid to policy-making and the ideal of scientist-as-advisor is not an actual reflection of what happens in practice (see also Yearley 2005: 164–5). For example, while it is assumed that scientific knowledge is developed autonomously and then becomes relevant to policy, scientists are increasingly investigating issues because of their policy relevance, even though policy questions do not tend to conform neatly to scientific disciplines or even across disciplines. Answering such questions scientifically can and does result in as much conflict as consensus. Again, the feasibility study is illustrative here, as there is some debate within the study itself, configured around the potential for rural development and climate change mitigation surrounding the use of organic agricultural practices. This conflict was largely due to contrasting ideologies within the research team, with some seeing the small-scale, organic approach as the way forward, while others suggested a separate strategy would be required and that there was little room for such practices from an industrial perspective (indicative of the aforementioned petro-chemical biases). Ultimately, agro-industrial approaches eclipsed other potentially beneficial (and more practical) approaches – a result that was almost inevitable, given the significant financial backing and wide spheres of influence of the various players in the agro-industrial lobby.

Evidence-making is only half of the equation, and we still need to examine the second part – policy-making – to identify how the governance apparatus uses the evidence, even if imperfect information is provided. Amid a range of contributing factors that comprise the complex dynamics and institutional structures through which policies are made and change (Keeley and Scoones 1999, 2003; see also Jones 2009), there are three broad explanations: first, that policy reflects political interests; second, that policy is a product of actor networks;

and third, that policy is a product of discourse. Detailed discussion of these is impossible here, but reframing policy analysis should consider all three and, in particular, nuances may be found at their intersections. The issue is that scientific knowledge or evidence is interpreted in policy-making by policy-makers (usually through the assistance of knowledge brokers and surrounding networks), who generally comprise both technical personnel and their political superiors. Even when the evidence is clear, its meaning for decision-makers may be ambiguous and is a function of interpretation.

Of particular importance is the increasing attention given to policy as sense-making, driven by storylines or policy narratives, which do the work of discourses and interpretive paradigms (Hodgson and Irving 2007). The importance of narratives is that they are an attempt to bring order to the complexity encountered when making policy. As Rein and Schön (1993: 146) highlight, narratives allow for a frame to be developed, which is a "perspective from which an amorphous, ill-defined, problematic situation can be made sense of and acted on." It is here where evidence may actually be gathered in response to existing political conceptualizations or problem definitions (see, for example, Sharman and Holmes 2010), rather than used to define the issue in the first place. That is, within the global biofuels discourse there is a powerful and easily digestible narrative in which producing biofuels through existing large-scale and industrial processes has positive benefits for both developed and developing countries, when the truth is far more complicated. Complex scenarios may prompt simplistic debate and limit the ability of those in opposition to formulate counter-positions in the face of overpowering interests (Lyall et al. 2009).

Governments may also be reluctant to engage in debate, even within the apparatus of the state. They may not consider the wider perspectives necessary to capture complexity once an issue has been framed by certain actors (usually specific departments or units). In such cases it may be very difficult to turn back and the outcome is unlikely to shift drastically even if new evidence suggests it would be apposite to do so. Hints of this can also be seen in the early development of the biofuels policy in South Africa, which started in earnest once the Directorate for Hydrocarbons and Energy Planning in the Department of Minerals and Energy (DME) began pushing the agenda forward. While a range of axes (one might call them policy streams) were evident prior to this juncture (largely involving the Department of Science and Technology and other stakeholders), the strategy's development within the DME shifted the focus to biofuels for transport and development along a modernization rationale, consistent with that department's perspective on energy use and the strategic focus on energy in South Africa.

This perspective opened biofuels up to criticism from, for example, advocates of small-scale, bioenergy perspectives and those who feared such approaches were contrary to the proposed benefits of job creation, climate change mitigation, and energy security. Without much financial backing and influence, these voices have not significantly affected the prevailing policy direction, apart from somewhat token rejections of maize (due to food security concerns) and jatropha (considered "alien" vegetation) as potential biofuels feedstock (South African Department of Minerals and Energy 2007). Both of these rejections were ill-considered, given that it is not the choice of crop so much as what is done with it that lies at the heart of the problem. Decisions taken at the national level have shackled potentially useful local projects by homogenizing complex and seemingly intractable issues such as food security into simplistic notions informed by inadequate understanding of causes and effects. However, with the 2008 food crisis emerging at the time when the policy was debated, it is easy to see why policy-makers made such decisions, especially given the significant evidence at their disposal (see Ruysenaar 2011 for detailed discussion).

Why is the focus on industrial fuels that are clearly devoid of many of the proposed benefits and, from emerging practices, unlikely to mitigate climate change to any significant degree? "Framing," given the emerging international discourse, is one explanation, but it should also be said that certain logics that are particular to South Africa underlie such a rationale. Governments thrive on managing and controlling, and to do this they need to be able to measure. Industrial biofuels fit this modus operandi nicely. Strictly controlled industrial production provides for a number of associated proxies. Targets in renewable energy, GHG reduction, and job creation, among others, are extrapolated from litres of biofuel produced, usually with such calculations used to gauge the "real" benefits derived. For example, it is far more difficult to measure the decline in GHGs when individual households shift from coal or kerosene to ethanol gel than it is to measure the output of a biofuel refinery. No matter how unrealistic they may be, biofuels figures still "make sense" and conform to the SMART (specific, measurable, attainable/achievable/allocated, realistic, and time-based) objectives of government.

Along with measuring, governments are predisposed to leveraging any management opportunities and identifying those options that are likely to provide higher returns. It is, for example, far easier to control tax regimes based on existing user systems, so industrial approaches are likely to be favoured. The support for industrial fuels is, then, also driven by the extant technological infrastructure with little need to transform existing technologies and surrounding legislation and operations significantly. Existing and well-established infrastructures are not easily rebuilt or replaced, and there is little need for such changes when industrial first-generation biofuels are adopted. However, it should be mentioned that, despite the limited physical changes required, the legislative changes in South Africa proved slow and difficult. The same would be true in most countries, depending on political will and bureaucratic efficiency.

Ultimately, the prevailing view is that the less chaotic the transformation or modification of the systems through which energy is produced and utilized, the better, regardless of any wider implications, be they for climate change or otherwise. Despite the wide range of proposed benefits, there is little room for manoeuvre when existing socio-technical systems define, to a large degree, how new technologies may or may not be used (Carolan 2010). In South Africa, the minerals–energy complex is considered a key factor here, with a sustained focus or received wisdom making cheap fossil-based energy (mostly coal) one of the defining features of the wider political economy (Fine and Rustomjee 1996) and dictating the form and function of the energy supply sector (Büscher 2009; Baker et al. 2014). So, while individual departments may define strategies and policies based on bounded rationalities, these are also a function of wider socio-technical assemblages, strategic (economy-wide) decision-making, and the existing political economy. Biofuel production is thus largely integrated within existing institutions rather than part of a broader exercise in reframing policy in light of the need to tackle climate change.

Conclusion

In the discussion presented above, we can see the importance of who has the power to frame issues within decision-making. However, the situation is more complicated than that. We have, for example, neglected the role of political oversight and have not been explicit about the exact networks and narratives underpinning specific directions of the biofuels policy developments in South Africa (especially those around the various biofuels projects that are forming a biofuels niche in that country). What is important is that much of the policy development has been underpinned by "win–win" narratives of job creation and agricultural development. Climate change has been largely neglected; it was simply not a political imperative at the time that the policy was developed and very little evidence existed to inform such a position anyway. However, even if such evidence had been available, it probably would have done little to sway the decisions and generally would have been interpreted through existing political viewpoints. Given the nature of the emerging biofuels projects in South Africa, it seems that climate change remains marginal, simply because it is bound to be on the broader political periphery of such an emerging economy.

Returning to our initial discussion, it is clear that policy-making can be inflected with the same techno-optimism that (initially at least) permeated and shaped much of the biofuels discourse. In so doing, specific frames and frameworks of understanding gloss over more complicated and complex interconnectivities that such policies need to govern. Feasibility studies fail to examine multi-level and cross-scale interactions not only because such interactions may be beyond the technical expertise of the researchers conducting such studies, but also because they do not neatly conform to existing broader conceptualizations and prescriptions. If we are to move beyond this analytical/conceptual double bind, far more attention needs to be paid to emphasizing and responding to the limitations of existing knowledge systems and how the political nature of governance systems may not respond to this new information in the ways that one might expect. Efforts to reframe other issues in light of climate change face similar challenges in that prior institutional practices and knowledge-based decision systems emerge to shape policies ostensibly related to the new framing, but carry with them institutional inertia that requires much more than reframing if substantive change of direction is to be accomplished.

References

Austin, G. (2006) "Review of the terms of reference in the publication entitled *An Investigation into the Feasibility of Establishing a Biofuels Industry in the Republic of South Africa*," Cape Town: AGAMA Energy.

Baker, L., Newell, P., and Phillips, J. (2014) "The political economy of energy transitions: The case of South Africa," *New Political Economy*, 19(6): 791–818.

Beddington, J. (2009) "Food security: A global challenge," paper presented at a BBSRC workshop on food security, London, 19 February.

Biofuels Task Team (2006) *An Investigation into the Feasibility of Establishing a Biofuels Industry in the Republic of South Africa*, Pretoria: Department of Minerals and Energy.

Büscher, B. (2009) "Connecting political economies of energy in South Africa," *Energy Policy*, 37(10): 3951–8.

Carolan, M.S. (2010) "Ethanol's most recent breakthrough in the United States: A case of socio-technical transition," *Technology in Society*, 32(2): 65–71.

Cash, D., Adger, W., Berkes, F., Garden, P., Lebel, L., Olsson, P., Pritchard, L., and Young, O. (2006) "Scale and cross-scale dynamics: Governance and information in a multilevel world," *Ecology and Society*, 11(2), 8.

Colebatch, H. (2009) "Governance as a conceptual development in the analysis of policy," *Critical Policy Studies*, 3(1): 58–67.

Collingridge, D. and Reeve, C. (1986) *Science Speaks to Power: The Role of Experts in Policy Making*, New York: St Martin's Press.

Coyle, W. (2007) "The future of biofuels: A global perspective," *Amber Waves*, US Department of Agriculture, www.ers.usda.gov/amber-waves/2007-november/the-future-of-biofuels-a-global-perspective.aspx#.VKqLoCvF9I4, accessed 5 January 2015.

Énergie, Environnement, Développement (ENDA) (2008) "Biofuels development in Africa: Illusion or sustainable development?," www.bioenergywiki.net/images/6/68/ENDA_Biofuels_Africa.pdf, accessed 6 January 2015.

Fargione, J., Hill, J., Tilman, D., Polasky, S., and Hawthorne, P. (2008) "Land clearing and the biofuel carbon debt," *Science*, 319(5867): 1235–8.

Fine, B. and Rustomjee, Z. (1996) *The Political Economy of South Africa: From Minerals–Energy Complex to Industrialisation*, Boulder, CO: Westview Press.

Fischer, F. (1990) *Technocracy and the Politics of Expertise*, London: Sage.

Food and Agriculture Organization (FAO) (2000) *The Energy and Agriculture Nexus*, Environment and Resources Working Paper No. 4, Rome: FAO.

Fortin, E. (2011) "Multi-stakeholder initiatives to regulate biofuels: The roundtable for sustainable biofuels," paper presented at the International Conference on Global Land Grabbing, www.iss.nl/fileadmin/ASSETS/iss/Documents/Conference_papers/LDPI/30_Elizabeth_Fortin.pdf, accessed 5 January 2015.

Hodgson, S.M. and Irving, Z. (2007) "Policy and its exploration," in S. Hodgson and Z. Irving (eds) *Policy Reconsidered: Meanings, Politics and Practices*, Bristol: Policy Press, pp. 1–18.

Jones, H. (2009) *Policy-Making as Discourse: A Review of Recent Knowledge-to-Policy Literature*, IKM Emergent–ODI Working Paper No. 5, Bonn: IKM Emergent Research Programme, European Association of Development Research and Training Institutes.

Keeley, J. and Scoones, I. (1999) *Understanding Environmental Policy Processes: A Review*, Sussex: Institute for Development Studies.
Keeley, J. and Scoones, I. (2003) *Understanding Environmental Policy Processes: Cases from Africa*, London: Earthscan.
Koning, N. and Mol, A.P.J. (2009) "Wanted: Institutions for balancing global food and energy markets," *Food Security*, 1: 291–303.
Lederer, E. (2007) "Production of biofuels is a 'crime,'" *Independent*, 27 October.
Levidow, L. (2013) "EU criteria for sustainable biofuels: Accounting for carbon, depoliticising plunder," *Geoforum*, 44(1): 211–23.
Lyall, C., Pappaioannou, T., and Smith, J. (2009) *Limits to Governance: The Challenge of Policymaking in the Life Sciences*, London: Ashgate.
Mol, A. (2010) "Environmental authorities and biofuel controversies," *Environmental Politics*, 19(1): 61–79.
Moore, D. (2008) "Biofuels are dead: Long live biofuels(?): Part one," *New Biotechnology*, 25(1): 6–12.
Perrow, C. (1999) *Normal Accidents: Living with High-Risk Technologies*, Princeton. NJ: Princeton University Press.
Philip, K. (2010) "Inequality and economic marginalisation: How the structure of the economy impacts on opportunities on the margins," *Law, Democracy and Development*, 14(1): 1–28.
Pielke, R.A. (2007) *The Honest Broker: Making Sense of Science in Policy and Politics*, Cambridge: Cambridge University Press.
Pimentel, D. (2004) "Ethanol fuels: Energy balance, economic and environmental impacts are negative," *Natural Resources Research*, 12(2): 127–34.
Rein, M. and Schön, D. (1993) "Reframing policy discourse," in F. Fisher and J. Forester (eds) *The Argumentative Turn in Policy Analysis and Planning*, Durham, NC: Duke University, pp. 145–66.
Royal Society (2008) *Sustainable Biofuels: Prospects and Challenges*, Policy Document 01/08, London: The Royal Society.
Ruysenaar, S. (2011) "Rethinking the food-versus-fuel debate: An appraisal of international perspectives and implications for the South African industrial biofuels strategy," *Environment and Society: Advances in Research*, 2(1): 124–48.
Schoneveld, G. (2011) *The Anatomy of Large-Scale Farmland Acquisitions in Sub-Saharan Africa*, Bogor: Centre for International Forestry Research.
Searchinger, T., Hamburg, S., Melillo, J., Chameides, W., Havlik, P., Kamen, D., Likens, G., Lubrowski, R., Obersteiner, M., Oppenheimer, M., Robertson, G., Schlesinger, W., and Tilman, G. (2009) "Fixing a critical climate accounting error," *Science*, 326(5952): 527–8.
Sharman, A. and Holmes, J. (2010) "Evidence-based policy or policy-based evidence gathering? Biofuels, the EU and the 10% target," *Environmental Policy and Governance*, 20(5): 309–21.
Smith, J. (2007) "Culturing development: Bananas, petri dishes and 'mad science' in Kenya," *Journal of Eastern African Studies*, 1(2): 212–33.
South African Department of Minerals and Energy (2007) *Biofuels Industrial Strategy of the Republic of South Africa*, Pretoria: Department of Minerals and Energy.
Stoker, G. (1998) "Governance as theory: Five propositions," *International Social Science Journal*, 50(155): 17–28.
van der Voet, E., Lifset, R., and Luo, L. (2010) "Life-cycle assessment of biofuels, convergence and divergence," *Biofuels*, 1(3): 435–49.

Von Braun, J. (2007) "When food makes fuel: The promises and challenges of biofuels," keynote address at the Crawford Fund Annual Conference, Canberra, 15 August.

Whittaker, J., Ludley, K., Rowe, R., Taylor, G., and Howard, D. (2010) "Sources of variability in greenhouse gas and energy balances for biofuel production: A systematic review," *Global Change Biology: Bioenergy*, 2(3): 99–112.

Yearley, S. (2005) *Making Sense of Science: Understanding the Social Study of Science*, London: Sage.

10
NOVEL FRAMINGS CREATE NEW, UNEXPECTED ALLIES FOR CLIMATE ACTIVISM

Andrew Szasz

Introduction: climate denial, policy paralysis

On 30 March 2014, the Intergovernmental Panel on Climate Change (IPCC) issued its latest summary of *Climate Change: Impacts, Adaptation, and Vulnerability* (IPCC 2014). The report warned of massive, disruptive societal impacts, some already being seen, soon to worsen dramatically. The very next day, on Washington's leading political commentary website, *The Hill*'s headline read, "UN climate report changes little on Hill" (Cama 2014). The Republican Senator James Inhofe was quoted as saying that "The IPCC report is another effort to scare people into believing in man-made global warming" (Cama 2014). Inhofe also called it "a distraction from real problems in the world." On Fox News, Bill O'Reilly warned that the United States could "destroy [its] economy ... embracing some kind of phantom global warming theory" (Fox News 2014).

This is our situation today. Scientists have their "hair on fire," to borrow a phrase used by Richard Clarke to describe analysts' rising concerns about an impending attack before 9/11, but nationally the political process remains paralysed, forced into stalemate by conservative "climate sceptics." The outcome of the 2014 elections, which gave the Republicans control of the Senate and therefore put climate deniers in the chairmanships of all of the relevant congressional committees, offers little hope of breaking the logjam. Opinion polls show the nation similarly polarized: large majorities of Democrats, clear majorities of independents, and a majority, even, of moderate Republicans all accept that climate change is real, that it is caused by human behaviour, and that it is serious; but only about a quarter of conservative Republicans agree (Leiserowitz et al. 2014).

Climate scientists and sociologists have tried, in various ways, to respond to the scepticism. Sociologists have identified the sources of funding that support climate denial or climate scepticism, tracing flows of money from the fossil fuel industry and from conservative billionaires to climate-denying think-tanks and foundations (Dunlap 2009; McCright and Dunlap 2000; Dunlap and McCright 2010; Brulle 2013). They describe not just the "who" but also the "how" of climate denial, the rhetoric and tactics employed to try to undermine understanding of, and therefore concern about, scientists' findings (McCright 2007; Freudenburg and Muselli 2010; McCright and Dunlap 2000).

Climate scientists have themselves tried, in their way, to answer climate denial. Their default strategy has been, simply, to continue to do more research and publish the findings. They are amassing ever more evidence, with the newer studies typically finding that change is happening faster than previously thought and that the impacts are going to be, if anything, even worse than depicted in their earlier studies. Taking a more aggressive tack, some scientists have tried to tarnish sceptics' reputations by showing that their "go-to" experts have no credibility as scientists. Steve Schneider, a leading climate scientist, for example, co-authored a paper in which the citation counts of climate scientists are compared to the citation counts of deniers (Anderegg et al. 2010). Climate scientists have published thousands of articles in peer-reviewed journals; the deniers' experts almost none. That may be compelling evidence for those who believe in the culture of peer-reviewed science, but it will not – indeed, it has not – persuaded the typical sceptic who thinks that scientists are a self-interested group willing to do whatever it takes, even falsify their data, to gain control of the scientific discourse, keep the exorbitant federal research funding flowing, and, like their liberal, environmentalist allies, accomplish their real objective: to justify the government's increasing, unmerited involvement in ordinary people's lives.

None of these strategies has had much success so far, in part because climate denial has been caught up in, and is today merely one aspect of, larger conservative political/ideological currents in US political culture. Climate, like abortion, gun control, immigration, and gay marriage, has become a front in the "culture wars" – a vehicle that has been used effectively, over the past couple of decades, to organize and mobilize the US political right wing.

I wish, here, to explore the potential for a different strategy. Instead of continuing to try to weaken scepticism through a process of debunking (i.e., showing who is really paying for the denialist campaign, or showing that denial is based on "data" that misrepresents what is really happening), it might be a better idea to broaden support for climate action by forging alliances with other actors in US society. Many of these might seem unlikely allies, but for their own reasons they are all becoming increasingly concerned about the tendency to do nothing.

How might such potential allies be identified? The obvious first criterion would be that a potential candidate has already started to reframe climate change as a legitimate concern or as a clear threat to their core interests. Applying that criterion, I have identified three such actors: the American military establishment and national security apparatus; the insurance industry; and major faith communities in the United States. All three have begun to reframe climate as a key matter that affects their institutions, and they are all beginning to work on climate issues in ways that might help to break the current policy logjam.

The chapter begins by documenting these novel reframings. Next, it shows why such reframings are a necessary but not yet sufficient basis for forging coalitions. Finally, the conclusion holds out the hope that these unlikely allies may yet join a powerful coalition to meet the climate change challenge.

Climate change as a threat to national security

By 2007, both the US military and the US national security apparatus were starting to frame global climate change as a national security issue and, therefore, as something that is directly relevant to their mission, and crucial to what is most important in terms of government priorities in Washington. Later documents show that, despite the rising political influence of the denialist movement, the military and the national security establishment have not wavered in expressing deep concerns about the future of US national security if climate change is left to play out without serious intervention. I begin with two early, exemplary texts: the CNA Corporation's *National Security and the Threat of Climate Change* (CNA Corporation 2007) (also discussed by Dalby, Chapter 6, this volume) and the National Intelligence Council's *National Intelligence Assessment* (Fingar 2008). These documents reveal that both the US Navy and one of country's key spy agencies are very concerned about the consequences of climate change.

Under the heading, "The Destabilizing Impacts of Climate Change," the CNA Corporation spells out the likely impacts of climate change for some of the most vulnerable nations of the world: reduced access to water, reductions in agricultural production, the spread of infectious diseases, coastal flooding, land loss, and displacement of population (CNA Corporation 2007: 13–16). The report then describes the social/political consequences: failed states, terrorism, mass migrations, and conflicts over scarce resources (CNA Corporation 2007: 16–18). It concludes:

> climate change poses a serious threat to America's national security ... extreme weather events, drought, flooding, sea level rise, retreating glaciers, habitat shifts, and the increased spread of life-threatening diseases ... [Such] natural and humanitarian disasters ... will likely foster political instability ... in many Asian, African, and Middle Eastern nations, causing widespread political instability and the likelihood of failed states ... internal conflicts, extremism.
>
> *(CNA Corporation 2007: 6)*

The single "takeaway" message of the report is: "Climate change acts as a *threat multiplier* for instability in some of the most volatile regions of the world" (CNA Corporation 2007: 6; emphasis added).

In 2007, Congress directed the National Intelligence Council (NIC), the federal government's leading national security agency, to include climate change in its next *National Intelligence Assessment* (Mazzetti 2007). Although the assessment itself remained classified, its findings were summarized for congressmen by Dr Thomas Fingar, Deputy Director of National Intelligence for Analysis and chairman of the NIC.

Fingar testified that "current scientific observations indicate the Earth's climate is changing" and "In some cases, changes ... are occurring faster and with larger magnitude than scientists anticipated as recently as ten years ago" (Fingar 2008: 5, 6). Citing the familiar list of likely impacts – food shortages; problems over access to water; increasing frequencies of floods, drought, and extreme weather events – Fingar declared: "We judge global climate change will have wide-ranging implications for US national security interests over the next 20 years" (Fingar 2008: 4). Climate change "will worsen existing problems – such as poverty, social tensions, environmental degradation, ineffectual leadership, and weak political institutions" (Fingar 2008: 4–5).

The military and the national security apparatus have continued to be steadfast in adhering to the "climate is a national security issue" frame ever since those early articulations. For instance, climate change featured prominently in the US Department of Defense's 2010 *Quadrennial Defense Review Report*. Representatives of the military establishment continued to go to Capitol Hill to testify about the gravity of the issue. Admiral Gunn, president of the American Security Project, talked of food and water shortages, the spread of tropical diseases, and weakened, failing states – a series of impacts that would create masses of climate refugees and "incubate extremism ... [C]limate change threatens unrest and extremism as competition for dwindling resources, especially water, spreads. Weak or poorly functioning governments will lose credibility and the support of citizens ... extremism will increasingly find willing recruits" (Gunn 2009: 1, 2). "Climate change," Gunn concluded, "poses a clear and present danger to the United States of America" (Gunn 2009: 4). Gunn's testimony was echoed by others: Jonathan Powers, chief operating officer of the Truman National Security Project (Powers 2009); Admiral Titley, the US Navy's oceanographer (Titley 2010b); Admiral McGinn of the CNA Corporation (McGinn 2010); and the retired General Wesley Clark (Clark 2010).

The national security apparatus has been similarly steadfast in expressing its concern (US Department of Defense 2014a, 2014b). In 2008, the NIC published *Global Scenarios to 2025*, a planning exercise in which teams of experts developed three possible scenarios for how world history may unfold between now and 2025. The first of these, "Borrowed Time," is a warning about what might happen if political leaders do not act in a timely manner:

> a world following a path that, without major changes, leads to an unsustainable future ... Governments [prove] incapable of finding creative solutions to newer problems (e.g., climate change, global terrorism). Short-term, stopgap

solutions to problems requiring a long-term commitment are ineffective. Lack of global leadership only worsens conditions ... The result is a world that is ill-equipped to deal with complex global dilemmas.

(NIC 2008: 9)

These same themes were restated the following year, when Dennis Blair, Director of National Intelligence, testified in front of the Senate Select Committee on Intelligence (Blair 2009). Blair told the committee that the intelligence community had focused increasingly on the issue over the past year:

IC [the intelligence community] judges climate change will have important and extensive implications for US national security interests over the next 20 years ... [Outside the US, climate change] will worsen existing problems such as poverty, social tensions, environmental degradation, ineffectual leadership, and weak political institutions ... threaten[ing] domestic stability in some states, ... particularly over access to increasingly scarce water resources. We judge economic migrants will perceive additional reasons to migrate because of harsh climates, both within nations and from disadvantaged to richer countries.

(Blair 2009: 42)

Furthermore, Blair predicted that climate change would impact US national security in other, less direct ways:

The United States depends on a smooth-functioning international system ensuring the flow of trade and market access to critical raw materials such as oil and gas, and security for its allies and partners. Climate change could affect all of these – domestic stability in a number of key states, ... access to raw materials, and the global economy more broadly – with significant geopolitical consequences.

(Blair 2009: 42–3)

The House of Representatives heard similar testimony from James Woolsey, former director of the US Central Intelligence Agency (CIA) (Woolsey 2009). In 2010, another NIC report, *Global Governance 2025: At a Critical Juncture*, reiterated the council's earlier analyses, and Rich Engel, the director of the NIC's Climate Change and State Stability Program, travelled around the nation giving speeches and doing PowerPoint presentations on potential problems faced by the United States (Engel 2010a, 2010b).

Climate change: a mortal threat to the insurance industry

With annual worldwide sales estimated at US$4.3 trillion – about 7 per cent of world GDP (Liedtke et al. 2009: 6) – insurance is the largest (or close to the largest) "industry" on earth. Not every segment of the insurance industry is equally vulnerable to climate change, but some important segments certainly are, particularly firms that insure against property loss or personal injury caused by severe weather events and the giant reinsurers that sit at the pinnacle of the industry. Given that this massive industry is extremely worried about climate change, the potential impact on US national politics is of a similarly large scale.

If big enough, a single extreme weather event has the power to bring any insurance company to its knees: "Following Hurricane Andrew ... the country's largest homeowner property insurer, State Farm Fire & Casualty, was brought to the brink of insolvency, necessitating a rescue by its parent (State Farm Group)' (Mills et al. 2001: 15). The industry's exposure to risk increases when extreme weather events are more frequent and/or when those events become more severe and cause greater damage (see also Grove's discussion, Chapter 11, this volume). Recent insurance publications feature charts that graphically show a marked trend towards more frequent and more costly weather-related events (Dlugolecki 2009: 3; Nutter 2013). As we know, climate change is predicted to increase both the frequency and the severity of such events.

Increasing premiums could be one answer for the insurance industry, but raising them high enough to match the greater – and ever-growing – risk might result in customers baulking at buying insurance at all: "In extreme cases, uncertainty will render a risk uninsurable [because it is] unable to be priced at a level palatable to customers ... [and if] costs are high enough, there will be no price point that satisfies both the insurer and the consumer, and there will be no market for the insurance, rendering the risk essentially uninsurable" (Hecht 2008: 1567, 1574). Worse, some observers point out that climate change could result in conditions that are so drastically different from past experience that "Current catastrophe models, epidemiological assessments, and litigation risk models are likely not adequate to predict future risk" (Hecht 2008: 1580). Hecht (2008: 1581) concludes, starkly, "Thus, many of the basic criteria of insurability are threatened by climate change."

Some insurers claim to be agnostic about the *cause* of climate change (Insurance Information Institute 2010; State Farm n.d.), but no large insurance corporation or insurance trade group denies that climate change is happening and that it is likely to have a major – and potentially disastrous – impact on them. The clearest and most categorical statements come from the reinsurance segment of the industry, such as Swiss Re, Zurich Re, Geneva Re, Lloyd's of London, and these companies' trade group, the Geneva Association. All of these firms sell protection to other insurance companies who want a hedge against getting into trouble if they are suddenly liable for pay-outs that they cannot cover. Reinsurers' warnings have been timed to coincide with annual meetings of the Conference of the Parties (COP) to the United Nations Framework Convention on Climate Change (UNFCCC) (Liedtke et al. 2009) and they feature prominently on the websites of individual reinsurance giants (Dlugolecki 2009; Geneva Association 2010). As a recent headline declared, there are "No climate-change deniers to be found in the reinsurance business" (Reguly 2013). Hecht (2008: 1585) observes, "The huge amount of risk held by a comparatively small number of reinsurers may explain why those companies have been the most proactive in addressing climate risks." The reinsurance sector not only completely "buys into" the science; it can sound downright radical, as when declaring that its ultimate goal is "to promote the successful conversion to a low-carbon economy and resilient society" (Liedtke et al. 2009: 5).

Rather than review numerous similar statements, websites, and white papers, I shall focus on one key testimony by the president of the Reinsurance Association of America, Franklin W. Nutter, made at a hearing of the US Senate Committee on Environment and Public Works on 18 July 2013. Nutter (2013) quotes Swiss Re:

> Today, global warming is a fact. Since the beginning of industrialization and the rapid growth of world population, man's activities – along with natural variability – have contributed to a change of climate manifesting itself as a considerable increase in global temperature. Climate change has the potential to develop into our planet's greatest environmental challenge of the twenty-first century.

He states why insurers should and do care:

> Property casualty insurers are more dependent on the vagaries of climate and weather than any other financial services sector. Within the insurance sector, reinsurers have the greatest financial stake ... The industry is at great financial peril if it does not understand global and regional climate impacts, variability and developing scientific assessment of a changing climate. Integrating this information into the insurance system is an essential function.
>
> Insurers see climate primarily through the prism of extreme natural events ... droughts, heatwaves, the frequency and intensity of tropical hurricanes, thunderstorms and convective events, rising sea levels and storm surge, more extreme precipitation events and flooding ... In the 1980s, the average number of natural catastrophes globally was four hundred events per year. In recent years, the average is one thousand. Munich Re's analysis suggests the increase is driven almost entirely by weather-related events. North America has seen a fivefold increase in the number of such events since 1980.
> *(Nutter 2013)*

Take, for example, the threat of storm damage for coastal communities:

> [T]here are 4.2 million homes along the Gulf and Atlantic coast exposed to storm surge ... One million of these are in the category of extreme risk to storm surge and another 839,000 in the high-risk category ... Twenty-three of the twenty-five most populous US counties are ocean-facing ...

the insured value of coastal properties (defined as replacement cost not market value) is expected to increase at a rate of 7 per cent per year, which means that values would double every decade. Together with changes in weather patterns, intensity, and number of events, the result, of course, is an inevitable rise in insured and uninsured damages globally and in the US.

(Nutter 2013)

Although he does not use the phrase, Nutter's testimony leaves no doubt that climate change is an existential threat to his industry.

Climate change and Christian framings

About 24 per cent of Americans are Roman Catholic; 15–18 per cent identify as adherents of one of the older, "mainline" Protestant churches (Methodist, Episcopal, Anglican, Presbyterian, Lutheran); and about 26 per cent identify as Evangelical (which includes the Southern Baptists, the second-largest faith community in the United States after Catholics). All told, then, about three-quarters of US citizens identify themselves as Christian. Although that number is lower than in earlier times, the United States is still overwhelmingly a Christian nation. Therefore, a Christian reframing of climate change could have a tremendous political impact.

Any religion that wishes to talk to its adherents about the environment, about climate, has to be able to articulate why the issue is of legitimate concern for people of faith. It cannot merely say, as a lay organization might, "We understand the science, so we understand that we must do something … and fast." In the conclusion to our book on religion and climate change, my co-authors and I argued that church leaders have to do "discursive work" to reinterpret sacred texts and generally held beliefs so that environmental concerns can be understood as issues that should and must be of concern to the faithful (Veldman et al. 2013). That is another way of saying that they have to reframe climate change.

As a general observation, applicable to the whole spectrum of Christian views, this reframing has been done in two ways. First, since nature is God's creation, it is sacred and God wants us to respect it, protect it, and not despoil it. Second, because Jesus clearly cared so deeply about the poor, and because we understand that all environmental degradation, including climate change, hurts the world's poor first and worst, Christians have a duty to care for the environment.

Although my students and I have gathered data on many faiths in the United States, here I limit my discussion to just three leading ones: the Roman Catholic Church; the United Methodist Church (on the climate issue, this and the other "mainline" Protestant faiths hold similar views); and the largest conservative Christian faith community, the Southern Baptists and Evangelicals.

When he was Pope, now Pope Emeritus Benedict spoke often about protecting the environment and about climate change (Catholic Climate Covenant n.d.b).

In his encyclical letter *Caritas in Veritate* (*Charity in Truth*), he wrote with great eloquence about Catholics' responsibility towards nature:

> The environment is God's gift to everyone, and in our use of it we have a responsibility towards the poor, towards future generations and towards humanity as a whole ... In nature, the believer recognizes the wonderful result of God's creative activity, which we may use responsibly to satisfy our legitimate needs, material or otherwise, while respecting the intrinsic balance of creation ... Nature ... has been given to us by God as the setting for our life. Nature speaks to us of the Creator (cf. Rom 1:20) and his love for humanity ... Nature is at our disposal not as "a heap of scattered refuse" but as a gift of the Creator who has given it an inbuilt order, enabling man to draw from it the principles needed in order "to till it and keep it" (Gen 2:15).
>
> *(Benedict XVI 2009)*

Pope Francis, Benedict's successor, has spoken out just as forcefully about the environmental crisis (Catholic Climate Covenant n.d.c). At the time of writing (late 2014), he was preparing a major statement on climate change, to be published in 2015 (Roewe 2014).

Catholic bishops in the United States began to address the environmental question back in 1991, when they published *Renewing the Earth: An Invitation to Reflection and Action on Environment in Light of Catholic Social Teaching* (US Conference of Catholic Bishops 1991). Ten years later, the bishops issued a pastoral letter specifically addressing the issue of climate change, *Global Climate Change: A Plea for Dialogue, Prudence and the Common Good* (US Conference of Catholic Bishops 2001), and in 2006 they authorized the formation of the Catholic Coalition on Climate Change (now called the Catholic Climate Covenant).

The covenant is supposed to serve "as a catalyst, convener and clearinghouse that urges Catholic individuals, families, parishes, schools and other organizations to embrace and act on Catholic teaching as it relates to care for creation and climate change" (Catholic Climate Covenant n.d.a). Its main activity, its "primary organizing tool," is the St Francis Pledge to Care for Creation and the Poor, and its website claims that "Over 10,000 Catholic individuals, families and parishes have taken the Pledge ... [D]ozens of Catholic dioceses, religious communities, and colleges/universities have also taken the Pledge" (Catholic Climate Covenant n.d.a).

The Catholic Church has also appointed 24 "climate ambassadors," distributed geographically around the nation, to visit individual churches and make presentations about the Church's views on climate change. These presentations cover the basics of climate science but also emphasize the ethical reasons for caring and action: "At its core, global climate change is ... about the future of God's creation and the one human family" (Catholic Climate Covenant n.d.d).

In loosely similar formulations, statements on the United Methodist Church (UMC)'s websites leave little doubt that the Church's leadership is passionate about the environment and the global climate:

> The earth lies polluted under its inhabitants; for they have transgressed laws, violated the statutes, broken the ever-lasting covenant ... (Isaiah 24, NRSV) ... The crisis facing God's earth is clear. We, as stewards, have failed to live into our responsibility to care for creation and have instead abused it in ways that now threaten life around the planet ... As a matter of stewardship and justice, Christians must take action now to reduce global warming pollution and stand in solidarity with our brothers and sisters around the world whose land, livelihood and lives are threatened by the global climate crisis.
> *(General Board of Church and Society of the United Methodist Church 2014)*

In 2008, the UMC's annual conference passed a resolution on global warming (General Board of Church and Society of the United Methodist Church 2008). This calls on all members of the "global church community ... to reduce human-related outputs of greenhouse gases." All Methodists are urged to

> learn about ... release of greenhouse gases and evaluate their own lifestyles to identify areas where reductions in production and release of greenhouse gases can be made ... to make their own congregations more aware of the issue ... [to] reduce greenhouse gas emissions from congregational infrastructure (church buildings, parsonages, vehicles, etc.) ... [and] to educate others outside their church communities on the need to take action on this issue.
> *(General Board of Church and Society of the United Methodist Church 2008)*

A year later, the UMC's Council of Bishops followed up with a pastoral letter, *God's Renewed Creation: Call to Hope and Action*. The letter begins:

> God's creation is in crisis. We, the Bishops of the United Methodist Church, cannot remain silent while God's people and God's planet suffer. This beautiful natural world is a loving gift from God ... God has entrusted its care to all of us, but we have turned our backs on God and our responsibilities. Our neglect, selfishness, and pride have fostered: pandemic poverty and disease; environmental degradation; and the proliferation of weapons and violence ... God calls us and equips us to respond.
> *(Council of Bishops of the United Methodist Church 2009)*

The bishops then directed all Methodist ministers to discuss *God's Renewed Creation* with their congregations. The "mainline" Protestant faiths are, today, largely politically liberal and they have no problem accepting findings from science, even when those findings conflict with literal interpretation of the scriptures.

It is much more surprising to find green theology and green activism at the conservative end of the Christian faith spectrum. Nevertheless, it is present there too, albeit not without controversy. As documented by Zaleha and Szasz (2013), green Evangelicals and environmentally minded Southern Baptists articulate a green "take" on their theology around concepts such as "creation care" and "stewardship." In 2006, the Evangelical Climate Initiative (ECI) issued an "Evangelical call to action." This laid out four "claims":

1   Human-induced climate change is real and increasing international instability.
2   The consequences of climate change will be significant, and will hit the poor the hardest.
3   Christian moral convictions demand our response to the climate change problem.
4   The need to act now is urgent.

(Zaleha and Szasz 2013: 209)

As early as 1970, likely influenced by the emergence of a modern environmental movement in the United States, the Southern Baptist Convention (SBC) passed a series of resolutions articulating its understanding of environmental "stewardship." The "Resolution on the environment" (1970) reads like a straightforward, sympathetic appropriation of the era's Earth Day environmentalism with a Christian veneer: "God has created man to be a creature who needs clean air, pure water, and an environment which contributes to his general health," but "man has created a crisis by polluting the air, poisoning the streams, and ravaging the soil" (SBC 1970). Pollution is the newest manifestation of man's Fall; we are so sinful that we are destroying the conditions upon which we depend for our own survival. However, the resolution ends on a positive note, urging "Christians everywhere to practice stewardship of the environment and to work with government, industry and others to correct the ravaging of the earth" (SBC 1970).

Another resolution, passed by the SBC four years later, is quite similar: "God ... is the author of the universe [and He] views His creation as being very good ... The Scripture confronts us with our responsibility to God as stewards." However, we have failed to conserve resources. We have been "selfish and nearsighted." Christians must "assume ... responsibilit[y]" (SBC 1974).

Although Evangelicals and Southern Baptists subsequently began to mirror conservative Republican concerns about the government going too far in terms of regulating business, and although they wished to distance themselves from an environmentalism that, to them, smacked of nature worship, the "creation care" and "stewardship" frames, once articulated, could not be simply suppressed and forgotten.

When the Evangelical Environmental Network published its "call to action" in 2006, a young Evangelical student was inspired to post a nearly identical document, modelled on the call – "The Southern Baptist declaration on environment and climate change" – on the internet (Merritt 2006). It attracted hundreds of endorsements. Although a backlash from the politically conservative leadership of the SBC soon thwarted that initiative, young Evangelicals have continued to engage in climate activism.

Framing, by itself, is not yet enough

Looking beyond official framing rhetoric and considering actual behaviour, it is clear that frames that converge do not automatically create conditions for political coordination or coalition. It might be said that convergent framings are necessary but not yet sufficient; impediments stand between coalition-facilitating rhetoric and the ability or willingness to forge alliances.

The military and the national security agencies have been quite active within the boundaries of what they are legitimately permitted to do. They have drafted numerous policy papers. They have frequently gone to Capitol Hill to testify in front of congressional committees. They have organized conferences on climate and security. Their spokesmen have travelled around the country to give PowerPoint presentations that show the connections between climate and security.

In 2009, the US Navy formed a task force to improve understanding of the implications of climate change for naval operations (Freeman 2009; Titley 2010a: 31) and the CIA opened a Center on Climate Change and National Security (CIA 2009). In 2010, the military started working with climate scientists to develop better climate models (Morello 2010), while the CIA revived a programme that fostered the sharing of data between the intelligence and environmental science communities (Broad 2010). The US Navy is actively exploring alternative, greener energy sources for its ships and facilities.

There is, then, explicit cooperation with the climate science community and de facto agreement with the climate movement community, but the latter is not likely to become more overt. The US Constitution and political culture substantially limit the military's and the security forces' participation in national politics, and rightly so. The military is engaged, but strictly within the bounds of its mission. The message is climate "adaptation" (US Department of Defense 2014a), which is far from directly engaging with the fundamental causes of climate change. Adaptation involves an awareness of how climate change will present challenges to the military's missions, and planning so that the military's "agility and preparedness" are not degraded (US Department of Defense 2014a: 5).

In the past decade and a half, the insurance industry has made some, albeit uneven, progress towards addressing the threat of climate change. Back in 2001, Mills et al. (2001: 8–9) reported:

> The words "Climate Change" stir anxieties and arouse controversies among insurers ... A few have taken definitive positions that there is a material

threat, ... [but] the vast majority of individual firms and most trade organizations have not indicated an opinion ... while others have adopted equally strong views to the contrary.

Eight years later, Mills (2009: 324) found substantial changes had taken place:

> Mainstream insurers have increasingly come to see climate change as a material risk to their business ... [In a survey of] insurance industry analysts around the world ... Climate change was rated number one [among] the top-10 risks facing the industry ... and most of the remaining 10 topics (e.g. catastrophic events ...) are also compounded by climate change.

Mills' survey of what insurance firms had been doing since the turn of the century identified "643 specific activities" related to climate change. That may sound impressive, but Mills (2009: 337) hastens to point out that those 643 activities "no doubt [represent] a tiny fraction of global policies, suggesting that the overall insurance market remains considerably underdeveloped in terms of climate change products and services."

Indeed, a different assessment, published by the consulting firm Ceres, criticizes

> this powerful industry's sluggish and uneven response to the ever-increasing ripples from global climate change ... [O]nly 11 of the 88 companies reported having formal climate risk management policies in place, and more than 60 percent of the respondents reported having no dedicated management approach for assessing climate risk.
>
> *(Leurig 2011: 3)*

The report concludes: "The survey responses paint a picture of an industry that, outside of a handful of the largest insurers, is taking only marginal steps to address an issue that poses clear threats to the industry's financial health" (Leurig 2011: 4).

Furthermore, when industry leaders get specific about the policy changes they would like to see, their wish-list is remarkably modest. During his testimony to the US Senate Committee on Environment and Public Works in 2013, Franklin Nutter listed the policy changes that the industry urged the federal government to implement: strengthening building codes; reforming the National Flood Insurance Program; strengthening the Coastal Barrier Resources Act;

> [r]equir[ing] the Army Corps of Engineers to assess climate risk for all projects; funding climate and weather research; ... [o]ffer[ing] tax credits and other incentives that will encourage individuals and communities to make choices and adopt behaviors that promise to be more resilient when threatened by extreme weather events.
>
> *(Nutter 2013: 20–1)*

Such measures are undeniably sensible, and they could be considered concrete steps towards the reinsurance industry's professed ultimate goal – "to promote the successful conversion to a low-carbon economy and resilient society" (Liedtke et al. 2009: 5) – but the overall tenor of the recommendations leaned towards piecemeal adaptation strategies rather than fundamental interventions.

This is only to be expected. Firms in the private sector have fiduciary responsibilities to their shareholders. Protecting short-term profits is not just good business; it is a legal obligation. So, although the industry looks ahead and can clearly see it will soon be in serious trouble, and although its rhetoric regarding climate change can sound quite radical, when it comes to concrete policy initiatives, it has stuck to advocating rather modest reforms that would merely offer individual firms some protection from exposure to climate risk.

It has been over ten years since the United States' Catholic bishops expressed their desire to have Catholics care about and do something about climate change. After seeing that not much had happened, the bishops eventually tried a more proactive approach, instructing those 24 "climate ambassadors" to visit churches across the nation. The results have been mixed. In some locales the reception has been positive, even enthusiastic.For instance, Catholics in San Jose, California, organized a sweeping and multifaceted "Green Initiative" (Ward 2009). However, even in that generally liberal community the reception was uneven, as my students and I learned when we interviewed the local climate ambassador. Several local churches were not particularly interested in seeing his PowerPoint presentation.

Confirming our emerging understanding that even a very strong climate message from the Pope or the country's bishops does not generate automatic agreement, a study published in November 2014 found that white US Catholics and white Evangelicals are equally sceptical about climate change. By contrast, Latino Catholics seem to be quite concerned about climate change (Jones et al. 2014). One can speculate about the reasons for this, but the evidence is clear; even when the Pope and the bishops declare that climate change is real and serious, that understanding has percolated down to the pews very unevenly.

The United Methodist Church backed up its declarations with substantial action. It has participated in interfaith coalitions. Its spokesmen have testified in congressional hearings. Along with its ecumenical partners, it has filed *amicus* briefs in Supreme Court cases. It has used its considerable wealth to invest globally in businesses and projects that aim to build resilience and a more sustainable society. Inside the Church, the Council of Bishops has directed every Methodist minister to carry the message contained in *God's Renewed Creation* to their churches. Every one of the Methodist ministers my students and I interviewed stated that they agreed with this pastoral letter and thought discussing it in their churches was a legitimate directive from the bishops. However, not all of them then did so. Those who thought their parishioners already agreed with or were largely ready to hear the message wrote and delivered sermons about it, organized study groups, and encouraged efforts to "green" their churches' buildings and grounds by becoming more energy efficient, lowering their carbon footprints, and replacing grass with drought-resistant plants that require less watering. In less hospitable communities, however, the ministers were worried that the environmental message would not be welcomed and that some parishioners would be alienated and move to another church down the street – not an idle concern, given that the Methodists have lost several million members in recent years. These ministers still insisted that they would bring up the environment and climate change *at some point*. But not yet; the time would have to be right.

Beyond specific concerns about alienating conservative churchgoers, I noticed another constraint when I attended several "mainline" Protestant churches' services. The parishioners had attended primarily to satisfy a social need – the need to feel part of a community, to be with like-minded people, to enjoy the familiarity and comfort of sharing rituals that they had known since childhood, to say the same prayers together, to sing the same songs together. To be clear, US churches do not hesitate to mobilize their memberships for some political causes (especially against birth control, abortion, and gay marriage in recent years). However, while I heard social and environmental issues mentioned in sermons, it was obvious that the people in the pews were not particularly interested in engaging with such topics.

In Roman Catholic and "mainline" Protestant churches, then, despite a strong commitment at the top to do something about climate change, the message has not fully filtered down to the pews. Meanwhile, in conservative Christianity, climate denial starts at the top. When Richard Cizik – the National Association of Evangelicals' Vice-President for Governmental Affairs, its top lobbyist in Washington, and a leader of its green faction – ran for the presidency of the organization, conservatives organized to defeat him. Similarly, efforts to pass a climate resolution at a Southern Baptist convention were defeated, and then attempts were made to isolate and intimidate the young minister who had championed that resolution (Zaleha and Szasz 2013).

Polls show that most conservative Christians are sceptical about climate change (Jones et al. 2014), but there are also signs of green activism among young Evangelicals. In April 2014, for example, a new group of young Evangelicals organized a "day of prayer and action" on the subject of climate change at Christian colleges across the country (Young Evangelicals for Climate Action 2014).

The reframings accomplished by the leaders of both the Roman Catholic Church and the "mainline" Protestant churches are impressive. The clashes between those who try to "green" the conservative Christian communities and those who oppose their efforts have been intense and dramatic. The results among the various faiths' grass roots have been uneven and apparently not very successful. Jones et al.'s (2014) recent study shows that some demographic segments within the Christian community agree with the scientists and the activists, but most do not. Clements et al.'s (2014) analysis of the "greening of Christianity" is based on two General Social Surveys, one from 1993 and the other from 2010. They find no evidence of "major changes in ... environmental concern" among "rank-and-file church members" between those two dates, and that "self-identified Christians are similarly less green than are non-Christians and nonreligious individuals in 1993 and 2010" (Clements et al. 2014: 388–9).

Concluding on a note of optimism

The armed forces and spies, big insurance firms, the dominant Christian faiths: environmentalists have rarely, almost never, thought of these powerful establishment institutions as potential allies. As I have shown, though, each of them has now reframed climate change as a threat to their core interests. For the armed forces and the spies, climate is a threat to national security; for insurance companies, it is a threat to their business models; for Christians, the frame is moral or theological, based on Christ's teaching that his followers should care for the poor or God's love for His creation (or both).

Such reframings have real promise. However, as I have also argued, there remain some serious impediments to concerted action. Nevertheless, there is potential for further development and at least de facto alliance-building.

Military and national security leaders have been speaking out. However, even though the US Constitution and the country's traditions stand in the way of them stepping more directly into the political process – and most Americans are deeply grateful that they live in a nation that, at least in principle, strictly limits what its military and security forces are allowed to do – they could speak louder. It was the US Army that stood up to Joe McCarthy in 1954, and the nation benefited greatly from it. Would the nation not similarly benefit if the US Navy stood up to denialist politicians like John Boehner, Ted Cruz, Marco Rubio, Mitch McConnell, and James Inhofe today?

Environmental and climate campaigners could, in turn, improve their outreach efforts to publicize the extent to which these core federal institutions agree with them. The US public seems more or less unaware that their military leaders are deeply worried about climate change. Given the public's deep respect for the military, better awareness of their views might prove to be a significant counterweight to ongoing denialist rhetoric.

The insurance industry knows that it is in harm's way. Up to now, it has advocated rather modest reforms; but that could change. As the impacts of climate change grow ever more severe and it becomes clear that the long-term viability of the whole industry is at risk, insurance firms and their trade associations may well see the need to push for more radical, systemic interventions. If and when it takes that step, this industry will be joined by others, and not just minor ones like ski resorts and winemakers, who are already suffering the impacts. Most notably, in 2014, we began to see some of the core institutions of the world's financial system – the World Bank, the International Monetary Fund, the World Economic Forum at Davos – expressing real urgency

(Deutsche Welle 2014; Elliott 2014; Environmental News Service 2014; Parry et al. 2014).

Finally, as environmental conditions worsen, the gap between leaders and adherents in the faith communities might begin to close. In times of great distress, people look en masse to such leaders for answers. As the situation worsens and Americans see more extreme weather events, rising food prices, increasing difficulty accessing clean, potable water, and more climate-driven social conflict elsewhere in the world, they might well turn to their religious leaders and heed what they are saying about the connection between climate change and human well-being.

Organized religion still has considerable moral authority in the United States (if not in other advanced industrialized nations). As the crisis deepens, faith communities may well become major players in national climate politics. As a possible analogy, consider the role played by the Southern Black churches in the Civil Rights Movement in terms of providing infrastructure, resources, and a clergy that had both leadership skills and tremendous moral authority.

As conditions worsen – and, unfortunately, they are likely to worsen – a "meta-frame" might well emerge: climate change as an existential threat to global human society. Such a framing has the potential to subsume and synthesize all other framings: security, economic, ethical, and theological imperatives. It might even prove sufficiently powerful to overcome the denialist frame that currently paralyses climate politics in the United States.

References

Anderegg, W., Prall, J., Harold, J., and Schneider, S. (2010) "Expert credibility in climate change," *Proceedings of the National Academy of Sciences*, 107: 12107–9.
Benedict XVI (2009) "Encyclical letter *Caritas in Veritate* of the Supreme Pontiff Benedict XVI to the bishops, priests and deacons, men and women religious, the lay faithful, and all people of good will on integral human development in charity and truth," www.vatican.va/holy_father/benedict_xvi/encyclicals/documents/hf_ben-xvi_enc_20090629_caritas-in-veritate_en.html, accessed 26 December 2014.
Blair, D. (2009) "Annual threat assessment of the intelligence community," testimony before the Senate Select Committee on Intelligence, 12 February.
Broad, W. (2010) "CIA revives data-sharing program with environmental scientists," *New York Times*, 4 January.
Brulle, R. (2013) "Institutionalizing delay: Foundation funding and the creation of US climate change counter-movement organizations," *Climate Change*, 122: 681–94.
Cama, T. (2014) "UN climate report changes little on Hill," *The Hill*, 31 March, http://thehill.com/blogs/e2-wire/202259-un-climate-report-brings-no-change-to-political-winds, accessed 22 December 2014.
Catholic Climate Covenant (n.d.a) "Catholic Climate Covenant: Overview," http://catholicclimatecovenant.org/about-us/, accessed 22 December 2014.
Catholic Climate Covenant (n.d.b) "Pope Emeritus Benedict XVI on climate change," http://catholicclimatecovenant.org/catholic-teachings/benedict-xvi/, accessed 22 December 2014.

Catholic Climate Covenant (n.d.c) "Pope Francis on care for creation," http://catholicclimatecovenant.org/catholic-teachings/pope-francis/, accessed 22 December 2014.

Catholic Climate Covenant (n.d.d) "US bishops," http://catholicclimatecovenant.org/catholic-teachings/bishops/, accessed 22 December 2014.

Central Intelligence Agency (CIA) (2009) "CIA opens Center on Climate Change and National Security," press release, 25 September.

Clark, General W. (2010) Testimony before the House Select Committee on Energy Independence and Global Warming, 1 December.

Clements, J., Xiao, C., and McCright, A. (2014) "An examination of the 'greening of Christianity' thesis among Americans, 1993–2010," *Journal for the Scientific Study of Religion*, 53: 373–91.

CNA Corporation (2007) *National Security and the Threat of Climate Change*, Alexandria, VA: CNA Corporation.

Council of Bishops of the United Methodist Church (2009) *God's Renewed Creation: Call to Hope and Action*, Lake Junaluska, NC: Council of Bishops of the United Methodist Church.

Deutsche Welle (2014) "Cost of climate change high on Davos agenda," 24 January, www.dw.de/cost-of-climate-change-high-on-davos-agenda/a-17385764, accessed 25 December 2014.

Dlugolecki, A. (2009) *The Climate Change Challenge*, Risk Management SC 1, Geneva: The Geneva Association.

Dunlap, R. (2009) "Why climate-change skepticism is so prevalent in the USA: The success of conservative think tanks in promoting skepticism via the media," *IOP Conference Series: Earth and Environmental Science*, 6(53), http://iopscience.iop.org/1755-1315/6/53/532010/pdf/ees9_6_532010.pdf, accessed 24 March 2014.

Dunlap, R. and McCright, A. (2010) "Climate change denial: Sources, actors, and strategies," in C. Lever-Tracy (ed.) *Routledge Handbook of Climate Change and Society*, London: Routledge, pp. 240–60.

Elliott, L. (2014) "Climate change will 'lead to battles for food,' says head of World Bank," *Guardian*, 3 April.

Engel, R. (2010a) "Climate change impact on national security," PowerPoint briefing presented at the Federation of Earth Science Information Partners meeting, Washington, DC, 5–7 January.

Engel, R. (2010b) "The impact of climate change on national security," *Proceedings on Climate and Energy: Imperatives for Future Naval Forces*, Climate and Energy Symposium 2010, Johns Hopkins Applied Physics Laboratory and CNA Corporation, 23–24 March, pp. 59–64.

Environmental News Service (2014) "World Bank backs carbon pricing, convenes leadership coalition," 22 September.

Fingar, T. (2008) "Statement for the record: National intelligence assessment on the national security implications of global climate change to 2030," testimony before the House of Representatives Permanent Select Committee on Intelligence, House Select Committee on Energy Independence and Global Warming, Washington, DC, 25 June.

Fox News (2014) "Bill O'Reilly: The truth about Obamacare and global warming," 31 March, www.foxnews.com/transcript/2014/04/01/bill-oreilly-truth-about-obamacare-and-global-warming/, accessed 19 January 2015.

Freeman, B. (2009) "Navy task force assesses changing climate," American Forces Press Service, 31 July.

Freudenburg, W. and Muselli, V. (2010) "Global warming estimates, media expectations, and the asymmetry of scientific challenge," *Global Environmental Change*, 20: 483–91.
General Board of Church and Society of the United Methodist Church (2008) "1031: Resolution on global warming," http://umc-gbcs.org/resolutions/resolution-on-global-warming-1031-2008-bor, accessed 22 December 2014.
General Board of Church and Society of the United Methodist Church (2014) "Climate justice," http://umc-gbcs.org/issues/climate-justice, accessed 22 December 2014.
Geneva Association (2010) "The insurance industry and climate change, contribution to the global debate, 2009," Geneva: The Geneva Association, www.munichre.com/en/group/focus/climate-change/index.html, accessed 26 July 2010.
Gunn, L. (2009) "Statement of Vice-Admiral Lee F. Gunn, USN (Ret.), President, American Security Project," testimony before the Senate Committee on Foreign Relations Hearing on Climate Change and Global Security: Challenges, Threats, and Global Opportunities, 21 July.
Hecht, S. (2008) "Climate change and the transformation of risk: Insurance matters," *UCLA Law Review*, 55: 1559–629.
Insurance Information Institute (2010) "Climate change: Insurance issues," http://iii.org/issue_updates/222678.html, accessed 9 April 2010.
Intergovernmental Panel on Climate Change (IPCC) (2014) *Climate Change 2014: Impacts, Adaptation, and Vulnerability*, http://www.ipcc.ch/report/ar5/wg2/, accessed 22 December 2014.
Jones, R., Cox, D., and Navarro-Rivera, J. (2014) "Believers, sympathizers, and skeptics: Why Americans are conflicted about climate change, environmental policy, and science," paper presented at the American Academy of Religion Annual Meeting, San Diego, 22 November.
Leiserowitz, A., Maibach, E., Roser-Renouf, C., Feinberg, G., and Rosenthal, S. (2014) *Politics & Global Warming, Spring, 2014*, Fairfax, VA: George Mason University Center for Climate Communication.
Leurig, S. (2011) *Climate Risk Disclosure by Insurers: Evaluating Insurer Responses to the NAIC Climate Disclosure Survey*, Boston, MA: Ceres.
Liedtke, P., Schanz, K., and Stahel, W. (2009) *Climate Change as a Major Risk Management Challenge: How to Engage the Global Insurance Industry*, Geneva: The Geneva Association.
McCright, A. (2007) "Dealing with climate change contrarians," in S. Moser and L. Dilling (eds) *Creating a Climate for Change: Communicating Climate Change and Facilitating Social Change*, Cambridge and New York: Cambridge University Press, pp. 200–12.
McCright, A. (2010) "Anti-reflexivity: The American conservative movement's success in undermining climate science and policy," *Theory, Culture & Society*, 27: 100–33.
McCright, A. and Dunlap, R. (2000) "Challenging global warming as a social problem: An analysis of the conservative movement's counter-claims," *Social Problems*, 47: 499–522.
McGinn, D. (2010) "Energy security, climate change, and national security," *Proceedings on Climate and Energy: Imperatives for Future Naval Forces*, Climate and Energy Symposium 2010, Johns Hopkins Applied Physics Laboratory and CNA Corporation, 23–24 March, pp. 375–83.
Mazzetti, M. (2007) "Spy chief backs study of impact of warming," *New York Times*, 12 May.
Merritt, J. (2006) "The Southern Baptist declaration on environment and climate change," http://www.baptistcreationcare.org/node/1, accessed 3 June 2015.
Mills, E. (2009) "A global review of insurance industry responses to climate change," *Geneva Papers*, 34: 323–59.

Mills, E., Lecomte, E., and Peara, A. (2001) *US Insurance Industry Perspectives on Global Climate Change*, Berkeley, CA: Lawrence Berkeley National Laboratory.

Morello, L. (2010) "Defense experts want more explicit climate models," *New York Times*, 24 June.

National Intelligence Council (NIC) (2008) *Global Scenarios to 2025*, Washington, DC: National Intelligence Council.

National Intelligence Council (NIC) (2010) *Global Governance 2025: At a Critical Juncture*, Washington, DC: National Intelligence Council.

Nutter, F. (2013) "Climate change: It's happening now," testimony before the Senate Committee on Environment and Public Works, 18 July, www.epw.senate.gov/public/index.cfm?FuseAction=Hearings.Hearing&Hearing_ID=cfe32378-96a4-81ed-9d0e-2618e6ddff46, accessed 23 March 2015.

Parry, I., Heine, D., Lis, E., and Li, S. (2014) *Getting Energy Prices Right: From Principle to Practice*, Washington, DC: International Monetary Fund.

Powers, J. (2009) "Written statement of Jonathan Powers, retired US Army captain, chief operating officer, Truman National Security Project," presented to Senate Committee on Environment and Public Works, Hearing on Climate Change and National Security, 30 July.

Reguly, E. (2013) "No climate-change deniers to be found in the reinsurance business," *Globe and Mail*, 28 November.

Roewe, B. (2014) "With Lima Accord reached, climate attention turns toward 2015," *National Catholic Reporter*, 16 December, http://ncronline.org/blogs/eco-catholic/lima-accord-reached-climate-attention-turns-toward-2015, accessed 25 December 2014.

Southern Baptist Convention (SBC) (1970) "Resolution on the environment," http://sbc.net/resolutions/amResolution.asp?ID=452, accessed 24 September 2012.

Southern Baptist Convention (SBC) (1974) "Resolution on stewardship of God's creation," http://sbc.net/resolutions/amResolution.asp?ID=453, accessed 24 September 2012.

State Farm (n.d.) "State Farm's position on climate change," www.statefarm.com/about/media/current/climate.asp, accessed 9 August 2010.

Titley, D. (2010a) "Global climate change," *Proceedings on Climate and Energy: Imperatives for Future Naval Forces*, Climate and Energy Symposium 2010, Johns Hopkins Applied Physics Laboratory and CNA Corporation, 23–24 March 2010, pp. 27–43.

Titley, D. (2010b) "The Navy's climate change interests: Statement of Rear Admiral David Titley, Oceanographer of the Navy, director, Task Force Climate Change," testimony before the House of Representatives, Committee on Science and Technology, Subcommittee on Energy and Environment, 17 November.

US Conference of Catholic Bishops (1991) *Renewing the Earth: An Invitation to Reflection and Action on Environment in Light of Catholic Social Teaching*, www.usccb.org/issues-and-action/human-life-and-dignity/environment/renewing-the-earth.cfm, accessed 22 December 2014.

US Conference of Catholic Bishops (2001) *Global Climate Change: A Plea for Dialogue, Prudence and the Common Good*, www.usccb.org/issues-and-action/human-life-and-dignity/environment/global-climate-change-a-plea-for-dialogue-prudence-and-the-common-good.cfm, accessed 22 December 2014.

US Department of Defense (2010) *Quadrennial Defense Review Report*, Washington, DC: Department of Defense.

US Department of Defense (2014a) *2014 Climate Change Adaptation Roadmap*, Alexandria, VA: Office of the Deputy Under Secretary of Defense for Installations and Environment.

US Department of Defense (2014b) *Quadrennial Defense Review 2014*, Washington, DC: Department of Defense.

Veldman, R., Szasz, A., and Haluza-DeLay, R. (2013) "Conclusion," in R. Veldman, A. Szasz, and R. Haluza-DeLay (eds) *How the World's Religions Are Responding to Climate Change: Social Scientific Investigations*, New York: Routledge, pp. 297–315.

Ward, R. (2009) "Catholic Green Initiative of Santa Clara County underway," *Valley Catholic*, 15 September, www.valleycatholiconline.com/viewnews.php?newsid=805&id=10, accessed 22 December 2014.

Woolsey, R.J. (2009) "Testimony of R. James Woolsey," testimony before the House of Representatives Committee on Energy and Commerce, Subcommittee on Energy and Environment Hearing on the Climate Crisis: National Security, Public Health, and Economic Threats, 12 February.

Young Evangelicals for Climate Action (2014) "'Day of Prayer for Climate Action 2014' makes headlines nationwide," 9 April, www.yecaction.org/2014/04/09/day-of-prayer-for-climate-action-2014-makes-headlines-nationwide/, accessed 26 December 2014.

Zaleha, B. and Szasz, A. (2013) "Keep Christianity brown! Climate denial on the Christian right in the United States," in R. Veldman, A. Szasz, and R. Haluza-DeLay (eds) *How the World's Religions Are Responding to Climate Change: Social Scientific Investigations*, New York: Routledge, pp. 209–28.

11
CATASTROPHE INSURANCE AND THE BIOPOLITICS OF CLIMATE CHANGE ADAPTATION

Kevin Grove

Introduction

For many scholars, the question of climate change politics appears to be self-evident; it involves struggles to create more equitable climate change mitigation and adaptation procedures, and more just adaptation outcomes (see Pelling 2010 for a detailed review). However, critical research increasingly indicates that this understanding of politics suffers from a limited political imaginary, a narrow framing of politics that constricts our understanding of what it is, how it can be carried out, what its goals and purposes should be, and who (or what) the subject of it is. This is a point Erik Swyngedouw (2009a, 2009b) raises in a pair of articles in which he argues that research on climate change has become "post-political." By employing this term, Swyngedouw is gesturing towards a tendency for seemingly political activities, such as implementing community-based resilience projects designed to empower marginalized peoples or developing translocal climate governance mechanisms, to become techno-managerial exercises in cultural engineering and socio-ecological control. Here, he draws on the political theory of Jacques Rancière (2010), who defines politics as the assertion of incommensurable difference. It is not about struggles to integrate excluded people more effectively into a given order of things, but rather the continued presence of that which lies beyond this series of inclusions and exclusions that structures this order. For Rancière, politics involves the irruption of difference whose presence destabilizes the series of identities and hierarchies that stabilize the foundations of socio-ecological order.

In this chapter, I utilize a case study of catastrophe insurance in the Caribbean to elucidate the stakes involved in different framings of climate change politics. Specifically, I analyse the development and effects of the Caribbean Catastrophic Risk Insurance Facility (CCRIF). This is a regional insurance facility that offers parametric insurance coverage to 16 member states. Parametric insurance does not cover actual losses; instead, it provides a predetermined remuneration amount once specific parameters (such as a hurricane's wind speed and distance from a specific measuring point) are met. The CCRIF is able to offer coverage for previously uninsurable catastrophe risks through a specific combination of catastrophe modelling and risk pooling. Catastrophe models simulate disaster events' impacts on state revenue, which allows the CCRIF to project the probability of specific loss events and price member states' catastrophe risks accordingly. Pooling diversifies these risks throughout the wider Caribbean region, which lowers the cost of coverage for individual states and enables them to transfer their risk to global reinsurance and capital markets. The CCRIF thus introduces a new tool to finance disaster recovery that links Caribbean governments with global markets around the problem of securing critical infrastructure against catastrophic risks (see Grove 2012).

From a conventional framing of climate change politics, catastrophe insurance products such as those offered by the CCRIF create new financial resources for vulnerable small island developing states. They thus help to address adaptation inequalities between developed and developing states (see, for example, Adger et al. 2006 and the special issue of *Climate Policy* 2006, on catastrophe insurance). However, in this paper, I frame the politics of the CCRIF through the analytical lens offered by biopolitics. This and its corollary biopower refer to a form of power concerned with securing, developing, regulating, and improving life itself, however this "life" may be understood – as a population of individual persons, as a complex socio-ecological system, or so forth (see Foucault 2003; Anderson 2012). In contrast to accounts of power couched in a modernist imaginary, which see power as a negative quantity exercised by an individual over other individuals, a biopolitical imaginary recognizes that power is a relational effect, and productive of the subjects that it seeks to regulate. As such, biopolitics enables researchers to analyse the power effects of seemingly apolitical interventions designed to improve the quality of life – such as catastrophe insurance programmes like the CCRIF. As I will demonstrate below, a biopolitical analysis works against the "post-politicizing" tendencies that dominate much research on climate change politics. Here, the term "post-political" signals how complex technocratic responses to climate change, couched in the language of averting catastrophe, foreclose properly political debate that might call into question the assumed naturalness of the prevailing social order (Swyngedouw 2009a; Rancière 2010). Rather than reducing climate change insecurities to a series of threats to be managed through better adaptation and resilience-building initiatives, biopolitics encourages us to ask *how* insecurities come to be framed in ways that force us to address them through techniques such as catastrophe insurance. It also enables us to unpack how these framings facilitate the formation of trans-boundary environmental governance regimes, networks of state and non-state institutions organized around shared understandings of the environment as a source of threat (see also Bulkeley 2005; Grove 2009). In short, a biopolitical framing of climate change politics allows us to recognize how climate change adaptation initiatives, such as catastrophe insurance, are never divorced from a wider field of power relations, but are instead important vectors along which these relations can be relayed and, potentially, transformed.

Following a brief discussion of the "post-politicization" of climate change politics, I analyse the CCRIF's catastrophe insurance through the framework that biopolitics offers. This helps us to recognize how the CCRIF enacts what I have called the "financialization of disaster management" (Grove 2012); it reconfigures disaster management around the rhythms of financial capital as well as the rationalities and imperatives of state security and capital accumulation in an uncertain global environment.

Against climate change (post-)politics

Critical researchers on disaster management and climate change politics have long recognized that adaptation is political on two counts. First, adaptation is impacted by formal political processes and institutions. Specifically, the ability of individuals or communities to adapt is limited by inflexible institutions ill-prepared for the challenges of adapting to dynamic socio-ecological change (Quarantelli 1998). Additionally, many of the determinants of adaptive capacity are themselves affected by governance, or the public–private networks through which policy is developed and implemented. For example, access to resources and knowledge, infrastructure, political relations, kinship relations, and a flexible institutional environment, to name but a few, shape and constrain the possibilities for adaptive action (Adger and Kelly 1999; Smit and Wandel 2006; Nelson et al. 2007). Second, along these lines, adaptation is also political because the process of adapting produces uneven outcomes in society (Eriksen and Lind 2009). Adaptations taken by some people in some regions may impact the adaptive capacities of others in unexpected ways, thus compounding inequalities and increasing vulnerabilities (Adger et al. 2006; Nelson et al. 2007). There are winners and losers in any adaptive action, but the game is usually rigged from the start; the people who adapt are usually those with the social, political, and financial resources to do so. Adaptation thus creates localized equity and justice issues surrounding the distribution of the costs and benefits of coping with climate change (Adger et al. 2006; Nelson et al. 2007). The impossibility of eliminating vulnerability opens up the question of whose vulnerabilities are acceptable and whose must be reduced (Nelson et al. 2007; see also Dalby 2002).

We can classify these two configurations of adaptation and politics as "representative" and "distributional" politics, respectively. If we analyse them from a biopolitical framing, we can see that they are linked through a limited view of the political, a common line of sight with its feet firmly planted in a modern understanding of sovereignty and politics. Sovereignty is conceptualized here as an expression of generally accepted norms, exercised through institutions of government that comprise autonomous and rational political subjects (after Mbembe 2005). Politics involves the collective process of constructing these norms through communication and negotiation. To the extent that there is struggle, it is a struggle of individual wills to shape social norms to benefit their interests, which is necessarily limited by other, equally sovereign, and competing wills. The exercise of sovereignty is thus a process of self-limitation that allows for the institutionalization of social norms. Political problems are reduced to issues of representation (the ability to have a place at the negotiating table) and participation (the ability to speak at this table). In these accounts of climate change politics, procedural and distributional inequalities and injustices can be rationally managed by allowing aggrieved parties a place and a voice in the setting of collective norms.

The modernist subjection of political problems to rational management is a hallmark of "post-politicization." In both climate change and disaster studies, it is particularly evident in the preponderance of attention given to issues of governance as a way to solve procedural and distributional injustices (Grove 2013c). This work stresses the need for democratically oriented governance arrangements that widen participation in adaptation activities, empower marginalized and vulnerable groups, and create more robust adaptation and resilience policies (Engle and Lemos 2010). It focuses attention on developing techniques – such as adaptive management, reflexive and participatory governance, collaborative learning, and participatory education practices – that can contribute to more inclusive and reflexive environmental governance regimes. This allows marginalized groups greater participation in the decision-making processes that affect their lives. It also increases their ability to influence how scarce resources can be used to mitigate the impacts of environmental change. As such, participatory and reflexive governance can thus address distributional injustices at the same time as it addresses procedural ones.

However, despite the good *intentions* motivating this style of research, critical scholars drawing on Foucauldian understandings of power and subjectivity have shown how this line of work often has unintended depoliticizing *effects*. For instance, critical resilience scholars have shown how participatory resilience programming creates passive, adaptable subjects who live with suffering, insecurity, and vulnerability rather than try to change their worlds (Reid 2012). The "empowered" subject is an effect of biopolitical assemblages that attempt to enfold progressively greater swathes of socio-ecological existence into governmental regulation (Grove 2013b). They are biopolitical not because they utilize calculatory techniques of biopolitics, such as statistics, but rather because they attempt to make "life" (however this is understood; see below) subject to governmental regulation.

This understanding of biopolitics offers a productive way of framing climate change and disaster politics in a way that works against these "post-politicizing" tendencies. Key here is how biopolitics reworks a number of commonplace assumptions about the nature of power and its operation (Grove 2013a). First, power is not a quantity that can be held and exercised by one individual over another. Rather than the ability of one person to impose their will on another weaker will, power is a relational effect. Specifically, it is enacted through techniques such as the confessional or surveillance, each of which attempts (in different ways) to monitor individual actions and bring them in line with certain ideal norms (Foucault 1995, 2007). It circulates between, for instance, the priest and the confessor (Foucault 2007); the guard in the panopticon and the prisoner (Foucault 1995); and the psychoanalyst and the analysand (Foucault 1990). These relations produce subjects whose behaviour needs to be regulated, surveilled, improved, and brought in line with certain norms (the sinful confessor, the misbehaving prisoner, the neurotic analysand), and subjects who can regulate the abnormal other (the priest, the guard, the psychoanalyst). This brings us to the second difference. For Foucault, power is not a negative force. It is not the negation of one will by a stronger one. Instead, it is productive; it produces objects of power, abnormal others requiring regulation, and subjects of power, who devise and implement mechanisms to transform conduct. Third, Foucault refers to this process of regulating the conduct of others as "government." Thus, government does not refer to institutional arrangements that exercise power; instead, it is the "conduct of conduct," to use his famous phrase (Foucault 2007), action on the action of others that produces power effects. Fourth, power is not the antithesis of freedom. Instead, power *relies* on freedom – the possibility for transgressing the limits of any socio-ecological order – as a continual source of provocations for new kinds of governmental intervention. The concept of biopolitics thus helps us to recognize that efforts to produce more intricate knowledge about climate change impacts, hazards, and vulnerability, among other things, and resilience and adaptation will necessarily have biopolitical effects. They produce new truths about socio-ecological relations that open these relations to governmental intervention and control (Grove 2014).

Framing climate change politics in terms of biopolitics thus replaces a post-political will to truth with the ethical and political considerations of what forms of life research on climate change politics attempt to secure. What forms of life are valued and deemed worthy of protection? What forms of life are rendered insecure and vulnerable, subject to the pressures and violence of governmental intervention, regulation, and control, all in the name of securing valued life? To illustrate how a biopolitical imaginary reframes climate change politics, the next section analyses the biopolitics of Caribbean catastrophe insurance.

Insurance and biopolitics

Researchers inspired by Foucault's work have demonstrated how insurance is a key technique of biopolitical regulation (see, for example, Ewald 1991; Collier 2008; Lobo-Guerrero 2010b). Insurance operates through statistical and actuarial calculations to turn an uncertain future into risks whose potential effects can be mitigated through the purchase of a policy. In the process, it creates risk-bearing subjects who come to see their lives in terms of risk. Risk charts a course of action in an uncertain world; it shows that the world is full of danger, but also identifies normative behaviours that will minimize that danger. In this way, insurance produces a specific form of life: people who sense, perceive, reflect on, and act in a world of risks that should be properly managed.

However, the temporality and spatiality of climate change impacts pose two problems for insurance. In terms of temporality, actuarial statistics construct probabilities of future events on the basis of their past occurrence. However, because climate change effects are non-linear, they cannot be predicted through actuarial calculations. In terms of spatiality, insurance commonly assumes loss events will be evenly distributed throughout a population or region; that is, the entire population will not be subjected to the loss event at once. Actuarial statistics cannot predict where and when losses will happen; they can only predict that, within a given period of time, and in a particular region, a given number of people within the entire population will suffer a loss. The assumption that a loss event will be evenly spread throughout a region is an important principle of insurability, since it allows insurers to set premiums at levels that allow the loss of particular individuals to be distributed among the entire population. However, catastrophic events such as droughts, floods, or intense hurricanes often affect entire regions. Because climate change impacts are unpredictable and affect entire populations, some scholars of risk and insurance – notably Ulrich Beck's (1992) "risk society" thesis – assert that stochastic events, such as climate change or terrorist attacks, are fundamentally uninsurable.

Against the assertions of uninsurability, scholars drawing on biopolitics have shown that insurance utilizes a different set of techniques to visualize unpredictable futures and price the risk of catastrophic events (Collier 2008; Aradau and van Munster 2011). A different kind of knowledge is in play here; rather than statistical-actuarial knowledge, catastrophic insurance operates through enactment-based forms of knowledge that quantify and value catastrophe risks by simulating multiple possible futures rather than extrapolating the future from the past. As a result, rather than the gradual withdrawal of insurance in the face of climate change, a number of new insurance techniques and products have emerged using enactment-based calculatory practices. Techniques such as weather derivatives, risk pooling, catastrophe risk modelling, and parametric insurance, to name a few, provide the insurance industry with new opportunities for accumulation in an uncertain environment (Johnson 2013).

The CCRIF is one example of an insurance facility that relies on these enactment-based techniques. To provide Caribbean states with quick access to capital in the wake of a disaster, the CCRIF uses parametric insurance coverage. This is different from traditional indemnity insurance in that it does not cover all losses, but rather indexes a predetermined remuneration amount to the occurrence of specific weather events (Lobo-Guerrero 2010b). In the case of the CCRIF's hurricane coverage, the relevant parameters are wind speed and distance from a particular measuring point. The price of parametric insurance is determined not by actuarial methods that calculate risk on the basis of past disaster occurrences, but by enactment-based forms of knowledge that calculate and price catastrophe risk through models and simulations. The CCRIF's coverage uses a standard five-layer catastrophe model to project the likelihood and resulting economic impact of disaster events – in this case, hurricanes with specific parameters (see CCRIF 2009). In brief, its models index state losses to wind speed and distance, use computer simulations to predict the probability of storms with specific parameters, and price the resulting catastrophe risk accordingly (Grove 2012).

These catastrophe modelling techniques, and their biopolitical implications, will be unpacked below. For now, we can note that a biopolitical framing encourages us to consider *how* catastrophe insurance produces governmental effects. There is nothing inherently political about techniques such as parametric insurance and risk pooling. However, they gain a political edge as they are taken up and deployed in specific contexts – that is, within a complex web of juxtaposed bodies (human and non-human) and the force relations that circulate between them (Foucault 1977). In this sense, biopolitics is less an answer to questions about power and more an incitement to unpack *how* techniques of power operate in specific times and places, and *how* they produce governmental effects (Dillon and Lobo-Guerrero 2008). These questions are the focus of the next section.

Contextualizing the CCRIF

The CCRIF is situated at the intersection of a variety of local and global processes. Its formation reflects one angle of what Luis Lobo-Guerrero (2010a) calls an "insurance imaginary" of climate change. This refers to discourses that frame climate change in such a way that insurance appears as a "natural," unproblematic solution to the challenges that climate change poses to conditions such as development and state security. By the term "discourse," I do not refer to the textual or linguistic content spoken by an individual subject. Discourse is not something that is mobilized by an independent, sovereign will. Instead, I am referring to the *system of relations* between different concepts that structure what count as true and false statements. So, rather than the content of speech, a discourse can be thought of as the rules governing speech. A key assumption here is that meaning is not inherent in concepts, but is instead a relational effect. Meaning results from the relations between different concepts – relations that are held in place through specific discourses. As such, a key concern with biopolitical framings of climate change politics is to identify these discourses by mapping out, in a sense, the relations between different concepts that enable actors to assert certain "truths" about what climate change is, what problems it presents, and how these problems might be addressed.

In terms of an insurance imaginary of climate change, climate change is presented as a problem because it threatens *development* (Grove 2010). Acute climate change impacts, such as flooding or hurricanes, threaten to destroy vital infrastructure systems, and thus undermine decades of development investment. Likewise, recovering from these events can force states to rely on international loans, increasing their debts and retarding development gains. *State security* becomes a matter of concern here as well, since the ability of states to provide basic protection and services to their populations is potentially undermined by catastrophic events. In the extreme, disaster losses may lead to state insolvency, which can precipitate social breakdown. However, climate change is a problem not only because of its impacts on development. It also creates new *risks* that may not be known by people in threatened regions. Without this knowledge of new risk landscapes, people may engage in development activities that unwittingly increase their exposure to future extreme climate events.

In these discourses, climate change is a problem because it threatens development and state security, and creates new forms of risk that may be unknown. According to insurance imaginaries of climate change, insurance offers a solution to each of these problems. It can provide remuneration for disaster losses and thus enable individuals, states, and society as a whole to adapt to climate change impacts in ways that would be impossible without financial compensation. It can also provide states with needed injections of capital following a disaster event. This can enable states to continue paying salaries and rebuild vital infrastructure that sustains the circulation of capital and state security forces. Perhaps most importantly, insurance also encourages people to become better risk managers. Identifying exposure to and reducing risk before a disaster event can lead to lower premiums and increased risk awareness.

Insurance imaginaries of climate change have circulated among the insurance industry, international development agencies, and development scholars since the late 1990s. For instance, a number of parametric insurance programmes for individual farmers were piloted across South Asia during the early 2000s. These sought to provide farmers with new financial tools to improve their adaptive capacities, while also creating new markets for the insurance industry (Skees 2008). However, a biopolitical framing of climate change politics recognizes that a discourse involves more than a system of relations between concepts. It also recognizes that these discourses are solidified through underlying power relations. This means that discourses will not be the same in every time and place. Instead, they "touch down" in place and produce their truth effects as they articulate with contextually specific force relations. This means that a discourse always embodies wider struggles within society. The production of truth is not confined to specific debates within a field of practice – such as disaster management, climate change studies, or insurance – but rather reflects the wider field of force relations in which these practices are always embedded (Foucault 1994).

This means that a biopolitical framing also focuses attention on the *intersection* between particular discourses and the context in which they are situated. In the Caribbean, insurance imaginaries of climate change articulated with a number of trajectories that were reconfiguring the significance of development, state security, disaster events, and insurance. While a number of factors contributed to these reconfigurations (see Grove 2013b), I will focus on three particularly significant moments: the region's financial exclusion from global insurance markets; post-Cold War transformations in economic development and state security; and a particularly devastating hurricane season in 2004.

First, the problem of catastrophic risk coverage has been a thorn in the side of Caribbean governments and insurers since a string of disaster events in the late 1980s and early 1990s (Poncelet 1997). These events made clear the deleterious impact that more frequent and intense hurricanes could have on the region's economies, as well as the region's financial exclusion (Leyshon and Thrift 1995) from global insurance markets. These conditions are detailed in a series of reports by Dennis Lalor, head of the Insurance Company of the West Indies, commissioned by the President of the Bahamas (see Grove 2012). In these reports, Lalor documents how the Caribbean had become one of the insurance industry's most high-risk regions, and how this had led major reinsurance firms in Europe and North America to scale back their investments in the region, in favour of larger and more lucrative US markets. The lack of reinsurance coverage meant that local insurers were unable to transfer their risks to global financial markets. This led the local firms to limit the amount of hazard and peril coverage that they provided and increase the cost of this coverage, which in turn left populations and businesses throughout the region either uninsured or underinsured. As a solution, Lalor proposed a regional catastrophe risk pool. This would be established through an initial investment by Caribbean governments and sustained through yearly premiums from local insurance providers. After a disaster event, insurers could collect out of the pool to cover their losses. The goal of risk pooling here was thus to create a self-sustaining catastrophe fund for regional insurance companies that diversified each company's risks throughout the region, rather than concentrating them within their domestic market. Lalor's proposed risk pool therefore did not attempt to address the *problem* of financial exclusion, but rather the *effects* of financial exclusion on the region's insurance companies.

Second, the end of the Cold War brought about changes in the global geopolitical economy that transformed the significance of both development and state security in the region. During the Cold War, the Caribbean had received preferential trading status from Western countries, and especially the United States. For instance, the US Caribbean Basin Initiative (CBI), a series of bilateral trade agreements signed during the 1980s, was designed to integrate Caribbean governments and US capital more deeply. Supporting macroeconomic development by giving Caribbean states preferential trading partner status was the "soft" side of US strategies to combat the spread of communism in the United States' "backyard" (Klak 2009). However, with the end of the Cold War, the economic linkages forged by the CBI and other multilateral trade deals, such as the Lomé Convention (signed with the European Commission) and the Caribbean–Canada Trade Agreement (CARIBCAN), increasingly became sources of threat and insecurity, especially as Caribbean ports became vital nodes in transnational drug trafficking networks (Griffin 2008; Ward 2008). At the same time, NAFTA and World Trade Organization rulings steadily undermined the preferential treatments that had been afforded to Caribbean manufacturers and farmers through these trade deals (Bernal 2000; Black and Kincaid 2001). The combination of growing dependence on international trade and an increasingly precarious ability to attract capital and maintain access to Western markets made Caribbean leaders more sensitive to issues that might negatively affect the region's investment climate. Key here were the disruptive possibilities of traditional concerns, such as organized crime and violence. However, leaders also acknowledged that the region's vulnerability to environmental hazards contributed to its marginalized standing in the global economy (Grove 2012). These concerns made the provision, maintenance, and security of critical infrastructure systems important matters not only of economic development but also of social stability and state security.

Third, the problems of financial exclusion, disasters, state security, and macroeconomic development converged in the particularly devastating 2004 hurricane season. The key event here was Hurricane Ivan, a Category 4 hurricane that hit Jamaica and caused losses of US$549 million, roughly 8 per cent of the country's GDP. The worst was yet to come, though. Ivan then passed directly over Grenada and inflicted US$2.4 billion of damage – more than double the country's GDP (Kambon 2005). This far exceeded the amount of damage that professionals had thought a Category 4 storm could cause in the Caribbean. Even more serious than the financial losses was the impact that Ivan had on the Grenadian state. A few years later, an official from the UK's Department for International Development declared: "The country was just completely decimated. They lost two hundred per cent of their GDP in one night. For a while, the state itself ceased to exist. It was housed on a British ship, trying to get things together" (interview with the author, 14 October 2009). The Grenadian state's existence was rendered precarious not only because of the havoc that Ivan wreaked on the island's infrastructure but because the storm's economic impact created a so-called "liquidity gap" that financially crippled the government, which had no access to capital and so had to delay relief and recovery efforts for several weeks (CCRIF 2008).

Although no social disorder resulted from this, the Grenadian state's near collapse because of a lack of liquidity raised the spectre of a frightening link between the immediate financial impact of hurricanes on state treasuries and the potential for social disorder lurking in the shadows. Jamaica's Minister of Finance at the time, Omar Davies, referred explicitly to this link when he told a conference of potential CCRIF donors organized by the World Bank:

> There are immediate short-term, medium-term, and long-term effects [from hurricanes]. These include the need for resources to pay workers, service debt, and effect repairs to critical social and economic infrastructure. *If some of these are not tackled quickly, there can be serious health and environmental risks and even social upheaval.*
>
> *(Davies 2007; emphasis added)*

As I have demonstrated elsewhere (Grove 2012), the Davies' understanding of the threat posed by disasters operates through a premediatory imaginary that seeks to envision a catastrophic scenario before it happens. That no such social upheaval occurred in Grenada following the state's collapse is immaterial. What matters is the *possibility*, not the probability, that a future catastrophe *might* trigger infrastructural breakdown and widespread social revolt.

Hurricane Ivan thus condensed a new imaginary of environmental security and development among Caribbean state leaders. The storm's impact folded together previously disconnected developments – the region's financial exclusion from global insurance markets, its newly precarious position in the global political economy, rising crime and violence, and more frequent and intense hurricanes – to reconfigure state security and development as problems of the state's access to financial capital in the immediate aftermath of a catastrophic event. The CCRIF emerged as a response to this particular vision of environmental insecurity. Its parametric coverage provides a mechanism for states to access capital in the wake of disaster and curtail the threats that disasters now pose to both state security and macroeconomic development. Specifically, the injection of capital offered by the CCRIF gives states the ability to, in the words of a World Bank document, "jumpstart recovery efforts quickly, limiting the liquidity impact on the government budget," which in turn allows them to focus on "providing direct assistance to the most affected citizens instead of working to raise funds from the international community to start the recovery" (World Bank 2005: 32).

From a biopolitical perspective, we can read the CCRIF as a specific governmental technology that responds to the particular problems that catastrophes pose to state security and development in the contemporary Caribbean. Its development and adoption are conditioned by wider transformations in the region's geopolitics and political economy. These transformations give the CCRIF's catastrophe insurance techniques their particular biopolitical edge. Although the CCRIF's activities are frequently couched in the language of humanitarian relief and progressive climate change adaptation, as above, its practical effects suggest that the CCRIF's catastrophe insurance coverage enacts a specific kind of biopower. The next section explores the CCRIF's biopolitical effects.

Financializing disaster management

Perhaps the most significant effect of the CCRIF is how it facilitates the entry of financial techniques and rationalities into disaster management. This is a process I have described as the "financialization of disaster management" (Grove 2012). "Financialization" refers to a political, economic, and cultural process of aligning the strategies, techniques, and space–times of disaster management with the rhythms and rationalities of global financial markets (Martin 2002). On the surface, it occurs through the CCRIF's catastrophe insurance products, which weave together member states, North American and European reinsurers, and global financial markets. Key here is the CCRIF's reinsurance structure.

The CCRIF sells catastrophe insurance coverage to member states but does not keep these risks itself. Instead, it transfers them to reinsurance and weather derivative markets. For example, in the 2008–9 hurricane season, the CCRIF purchased US$132.5 million in reinsurance spread across three layers. The first layer, which covered one-in-seven-year events, cost US$12.5 million. A US$30 million layer covered one-in-twenty-five-year losses. The top layer – US$90 million – extended the CCRIF's claims-paying capacity to one-in-fourteen-hundred-year loss events. Thirty million dollars of this top layer were placed into global capital markets through weather derivatives issued by the CCRIF and sold to the World Bank Treasury (Grove 2012). This reinsurance gave the CCRIF a claims-paying capacity exceeding one-in-ten-thousand-year loss events, effectively ensuring that it would remain solvent in virtually any loss event.

However, from a biopolitical angle, the ability of the CCRIF to link states with reinsurance and capital markets and provide them with post-catastrophe remuneration is of less importance than *how* and *to what effect* this state of affairs is realized. Key here are the CCRIF's catastrophe models. These function as cogs in the machinery of catastrophe insurance. Specifically, they determine the levels of state losses that constitute loss events with specific probability and price these risks accordingly. The CCRIF uses a standard five-layer catastrophe model to project state losses, as damage to the built environment, from storm events with specific parameters. The first layer is the hazard module, which uses 150-year data to model a particular event's effects, such as a storm with a specific wind speed and distance from a given measuring point (such as a local airport). The second layer is the exposure module, which determines the value of the built environment exposed to the particular peril. The CCRIF's models use satellite imagery to estimate building and infrastructure cover, and its replacement cost, within 900-square-metre grids. This provides a fine-grained representation of each island's exposure. Third, the vulnerability module estimates the mean damage ratio (MDR) – the percentage of a structure's replacement costs that will be required to repair it after an event of specific magnitude. The MDR increases along with the intensity of the storm event, so a Category 5 storm ten kilometres from the island may have an MDR of 0.90, while a Category 3 storm 20 kilometres away may have an MDR of 0.50. The fourth level, the damage module, calculates projected losses by multiplying the MDR for each asset class by the value at risk determined in the exposure module. The fifth module, the loss module, aggregates the damages for each asset class in order to determine the overall economic impact of an event on the state's revenue (Grove 2012). This model is then run through thousands of simulations to generate probability functions for specific loss levels.

Because the CCRIF's probability calculations are produced by simulating future disaster events, these models are a form of enactment-based knowledge about member states' catastrophe risk. These calculations become the basis for the parametric insurance contract that member states purchase from the CCRIF. In essence, catastrophe modelling turns place-based uncertainties – the unknown effect that a future disaster will have on the built environment and critical infrastructure of an island – into an abstract, mobile, comparable, and commodifiable quantity – catastrophe risk – that can be sold, bought, divided, and resold to global reinsurance and financial markets through parametric insurance products, reinsurance contracts, and weather derivatives.

However, the calculation and commodification of catastrophe risk does not encompass the entirety of disaster management's financialization. The abstraction, commodification, and transfer of catastrophe risk also reconfigure the underlying rationalities and space–times of disaster response. This situation and its biopolitical effects were indirectly signalled to me during an interview with a director of a national disaster management agency whose country had recently received a parametric insurance payout from the CCRIF. The director explained his frustrations with the CCRIF's payout in the following terms:

> In our case, the money that was supposed to come from the CCRIF was earmarked for refurbishment of government buildings that were damaged. Come check [our island]: the government buildings are still damaged but the money's gone. Where it went was the general funds, so when stuff needed to be paid at that time, it covered those expenses. It's a case where, OK, this money's been set aside but you have bills to pay and you put it in the general pool where all the government utilizes it. Then we don't know how much of that money went to what.
>
> *(interview with the author, 9 December 2009)*

The director's frustrations help us to draw out two significant effects of catastrophe insurance. First, the CCRIF enables states to direct the post-disaster flow of capital in new ways that leave their mark on the landscape. Because the CCRIF often signs parametric insurance contracts with ministries of finance, not disaster management agencies, the payouts go directly to those ministries, not to disaster management professionals. As a result, in this case, funding that should have gone to repairing damaged buildings was channelled instead to unknown purposes, leaving a landscape still scarred by the disaster event over a year later. Here, the CCRIF's insurance contract effectively reconfigured the flow of disaster recovery financing away from repairing the damaged environment and towards the state's new environmental security imperatives, which may have included repairing some of the built environment but probably also entailed debt servicing or filling a general liquidity gap.[1] The still-damaged landscape was one material expression of financialized disaster management's new calculus of value and importance.

Second, reflecting on the reconfiguration of financialized disaster management evident in landscapes of disrepair can help us to recognize how catastrophe insurance alters the meaning and value of people affected by catastrophic events (Lobo-Guerrero 2010b; Grove 2012). No longer are they individuals coping with physical, economic, and emotional trauma, in need of relief and assistance to rebuild their damaged lives with dignity. Instead, they increasingly become objects of state security: a faceless population of potential threats to social stability and state security. In financialized disaster management, critical infrastructure that is repaired is valued not for its ability to promote human livelihoods and well-being, but rather for the role that it plays in maintaining a docile and ordered population that will not threaten foreign investment. This is catastrophe insurance's ultimate biopolitical effect. Financialized disaster management is not organized around the principle of alleviating the hardship and indignity that disasters cause; instead, it utilizes techniques such as catastrophe insurance to pre-empt and negate the threat that suffering humans pose to state security in a globally interconnected political economy.

Conclusions

Through the example of catastrophe insurance in the Caribbean, this chapter has sought to illustrate the utility of a biopolitical framing of climate change politics. This does not approach catastrophe insurance as a solution to problems of inequality or a lack of adaptive capacity simply in need of fine-tuning but rather unpacks the contextually specific ways in which catastrophe insurance and its associated techniques – such as parametric insurance, risk pooling, and catastrophe modelling – create objects and subjects of governmental regulation. In the case of the CCRIF's catastrophe insurance, the CCRIF is a key cog in the ongoing financialization of disaster management. Financialized disaster management produces new possibilities for accumulation, as catastrophe modelling and risk pooling make state risk in the Caribbean "a good bet" for European and North American reinsurers (Grove 2012). In the process, it also reconfigures disaster management around the imperatives, rationalities, and time–spaces of global financial circulation and Caribbean-specific understandings of state security. Financialization directs flows of disaster response resources away from repairing damaged buildings and providing relief supplies to suffering peoples and towards the post-disaster maintenance of critical infrastructure, the control and regulation of affected populations, and the prevention of liquidity gaps.

As this example demonstrates, a biopolitical framing enables a critical analysis of adaptation and resilience-building techniques that works against the post-political turn in many strands of climate change and disaster research. We can briefly draw out two key benefits here. First, a biopolitical framing recognizes that mitigation, adaptation, and resilience-building initiatives may be designed and deployed with the best of intentions, but those intentions are independent of the governmental effects they may generate (see also Ferguson 1994 on well-intentioned development programmes). Thus, it is less concerned with questions of a technical nature – "How can we improve this community's adaptive capacity? How can we empower vulnerable populations? How can we transition to more resilient futures?" – than with interrogating the conditions of possibility that enable these technical questions to be posed in the first place. This is a key analytical realignment, for it avoids the liberal will to truth that seeks to fold everything from the environment and social relations to change itself into a calculated, objective form of knowledge. Rather than proceeding down a path that leads to the "post-politicization" of climate change politics, a biopolitical framing constantly forces researchers and activists to confront and consider the partial nature of "solutions" to problems of inequality and maldistribution, and the unintended forms of categorical violence that these "solutions," such as catastrophe insurance, may generate.

Second, along these lines, a biopolitical framing radically opens up the question of security in climate change politics. Security, of one sort or another, is frequently invoked in calls to avoid "dangerous" climate change, to adapt to its impacts, and to become resilient (see Grove 2010, 2014). However, these invocations of security pass over the key political question: security for whom or what (Dalby 2002, 2009)? What kind of community is threatened by the dangers of climate change? What forms of life should adapt and why? These questions and others are not asked in the "post-political" rush to prevent, prepare for, and adapt to climate change. In contrast to this technical focus, a biopolitical framing makes climate change adaptation an ethico-aesthetic and political problem. The insecurities that we now feel can provoke us to invent new forms of life, acts of aesthetic production that enable new forms of encounter not only with the non-human environment but also with other humans. However, repeated calls for security and resilience channel this constitutive unease towards technical solutions that merely secure the institutions of liberalism, representative democracy, and capitalist markets that have created these insecurities in the first place. The political value of a biopolitical framing of climate change adaptation lies in its affirmation of the possibility that other forms of life are possible through the act of critique.

Note

1 This is not to place blame or responsibility for this configuration on the CCRIF, which makes a point of not getting involved in intragovernmental debates over which agency controls potential parametric payouts. The point here is merely to note the new financialized reality that confronts many disaster management professionals.

References

Adger, W.N. and Kelly, P.M. (1999) "Social vulnerability to climate change and the architecture of entitlements," *Mitigation and Adaptation Strategies for Global Change*, 4(3–4): 253–66.

Adger, W.N., Paavola, J., Huq, S., and Mace, M.J. (2006) *Fairness in Adaptation to Climate Change*, Cambridge, MA: MIT Press.

Anderson, B. (2012) "Affect and biopower: Towards a politics of life," *Transactions of the Institute of British Geographers*, NS 37(1): 28–43.

Aradau, C. and van Munster, R. (2011) *Politics of Catastrophe: Genealogies of the Unknown*, London: Routledge.

Beck, U. (1992) *Risk Society: Towards a New Modernity*, London: Sage.

Bernal, R. (2000) "The case for NAFTA parity for CBI countries," in K. Hall (ed.) *The Caribbean Community: Beyond Survival*, Kingston: Ian Randle, pp. 492–503.

Black, S. and Kincaid, J. (2001) *Life and Debt*, New York: New Yorker Video.

Bulkeley, H. (2005) "Reconfiguring environmental governance: Towards a politics of scales and networks," *Political Geography*, 24(8): 875–902.

Caribbean Catastrophic Risk Insurance Facility (CCRIF) (2008) *Annual Report 2007–2008*, Grand Cayman: CCRIF.

Caribbean Catastrophic Risk Insurance Facility (CCRIF) (2009) *Annual Report 2008–2009*, Grand Cayman: CCRIF.

Climate Policy (2006) Special issue: "Climate Change and Insurance," 6(6): 599–684.

Collier, S. (2008) "Enacting catastrophe: Preparedness, insurance, budgetary rationalization," *Economy and Society*, 37(2): 224–50.

Dalby, S. (2002) *Environmental Security*, Minneapolis, MN: University of Minnesota Press.

Dalby, S. (2009) *Security and Environmental Change*, Cambridge: Polity.

Davies, O. (2007) Presentation at the Donors' Pledging Conference on the Caribbean Catastrophic Risk Insurance Facility, Washington, DC, 26 February.

Dillon, M. and Lobo-Guerrero, L. (2008) "Biopolitics of security in the 21st century: An introduction," *Review of International Studies*, 34(2): 265–92.

Engle, N. and Lemos, M.C. (2010) "Unpacking governance: Building adaptive capacity to climate change of river basins in Brazil," *Global Environmental Change*, 20(1): 4–13.

Eriksen, S. and Lind, J. (2009) "Adaptation as a political process: Adjusting to drought and conflict in Kenya's drylands," *Environmental Management*, 43(5): 817–35.

Ewald, F. (1991) "Insurance and risk," in G. Burchell, C. Gordon, and P. Miller (eds) *The Foucault Effect: Studies in Governmentality*, Chicago, IL: University of Chicago Press, pp. 197–210.

Ferguson, J. (1994) *The Anti-Politics Machine: "Development," Depoliticization, and Bureaucratic Power in Lesotho*, Minneapolis, MN: University of Minnesota Press.

Foucault, M. (1977) "Theatrum philisophicum," in D. Bouchard (ed.) *Language, Counter-Memory, Practice: Selected Essays and Interviews*, Ithaca, NY: Cornell University Press, pp. 165–97.

Foucault, M. (1990) *The History of Sexuality, Volume 1: An Introduction*, New York: Vintage.

Foucault, M. (1994) *The Birth of the Clinic*, New York: Vintage.

Foucault, M. (1995) *Discipline and Punish: The Birth of the Prison*, New York: Vintage.

Foucault, M. (2003) *"Society Must Be Defended": Lectures at the Collège de France, 1975–1976*, New York: Picador.

Foucault, M. (2007) *Security, Territory, Population: Lectures at the Collège de France, 1977–1978*, New York: Picador.

Griffin, I. (2008) "A new conceptual approach to Caribbean security," in K. Hall and M. Chuck-A-Sang (eds) *The Caribbean Community in Transition*, Kingston: Ian Randle, pp. 221–40.

Grove, K. (2009) "Rethinking the nature of urban environmental politics: Security, subjectivity, and the non-human," *Geoforum*, 40(2): 207–16.

Grove, K. (2010) "Insuring 'our common future'? Dangerous climate change and the biopolitics of environmental security," *Geopolitics*, 15(3): 536–63.

Grove, K. (2012) "Preempting the next disaster: Catastrophe insurance and the financialization of disaster management," *Security Dialogue*, 43(2): 139–55.

Grove, K. (2013a) "Biopolitics," in C. Death (ed.) *Critical Environmental Politics*, London: Routledge, pp. 22–30.

Grove, K. (2013b) "From emergency management to managing emergence: A genealogy of disaster management in Jamaica," *Annals of the Association of American Geographers*, 103(3): 570–88.

Grove, K. (2013c) "Hidden transcripts of resilience: Power and politics in Jamaican disaster management," *Resilience*, 1(3): 193–209.

Grove, K. (2014) "Biopolitics and adaptation: Governing social and ecological contingency through climate change and disaster studies," *Geography Compass*, 8(3): 198–210.

Johnson, L. (2013) "Catastrophe bonds and financial risk: Securing capital and rule through contingency," *Geoforum*, 45: 30–40.

Kambon, A. (2005) "Caribbean small states' development and vulnerability," *ECLAC Caribbean Development Review*, 1: 104–39.

Klak, T. (2009) "Development policy drift in Central America and the Caribbean," *Singapore Journal of Tropical Geography*, 30(1): 18–23.

Leyshon, A. and Thrift, N. (1995) "Geographies of financial exclusion: Financial abandonment in the United States and Great Britain," *Transactions of the Institute of British Geographers*, NS 20(3): 312–41.

Lobo-Guerrero, L. (2010a) "Insurance, climate change, and the creation of geographies of uncertainty in the Indian Ocean region," *Journal of the Indian Ocean Region*, 6(2): 239–51.

Lobo-Guerrero, L. (2010b) *Insuring Security: Biopolitics, Security, and Risk*, London: Routledge.

Martin, R. (2002) *The Financialization of Daily Life*, Philadelphia, PA: Temple University Press.

Mbembe, A. (2005) "Necropolitics," *Public Culture*, 15(1): 11–40.

Nelson, D., Adger, W.N., and Brown, K. (2007) "Adaptation to environmental change: Contributions of a resilience framework," *Annual Review of Environment and Resources*, 32: 395–419.

Pelling, M. (2010) *Adaptation to Climate Change: From Resilience to Transformation*, London: Routledge.

Poncelet, J.L. (1997) "Disaster management in the Caribbean," *Disasters*, 21(3): 267–79.

Quarantelli, E.L. (1998) "Disaster planning, emergency management, and civil protection: The historical development and current characteristics of organized efforts to prevent and to respond to disasters," DRC Preliminary Paper No. 228, Newark: University of Delaware Disaster Research Center.

Rancière, J. (2010) *Dissensus: On Politics and Aesthetics*, London: Bloomsbury.

Reid, J. (2012) "The disastrous and politically debased subject of resilience," *Development Dialogue*, 58: 67–79.

Skees, J. (2008) "Innovations in index insurance for the poor in lower income countries," *Agricultural Resources and Economics Review*, 37(1): 1–15.

Smit, B. and Wandel, J. (2006) "Adaptation, adaptive capacity and vulnerability," *Global Environmental Change*, 16: 282–92.

Swyngedouw, E. (2009a) "Apocalypse forever? Post-political populism and the spectre of climate change," *Theory, Culture & Society*, 27(2–3): 213–32.

Swyngedouw, E. (2009b) "The antinomies of the postpolitical city: In search of a democratic politics of environmental production," *International Journal of Urban and Regional Research*, 33(3): 601–20.

Ward, C. (2008) "The security–development nexus in US–Caribbean relations," in K. Hall and M. Chuck-A-Sang (eds) *The Caribbean Community in Transition*, Kingston: Ian Randle, pp. 136–55.

World Bank (2005) "Project appraisal document on a proposed grant in the amount of SDR 6 million (US$9 million equivalent) to the Republic of Haiti for a Haiti catastrophe insurance project," World Bank Report No. 38540-HT, Washington, DC: World Bank.

12
RESISTING THE CLIMATE SECURITY DISCOURSE

Restoring "the political" in climate change politics

Angela Oels

Introduction: climate change-induced migration as a security threat

Some years ago, there was a competition in the media to identify "the first climate refugees." For the Global South, the relocation of 1,500 residents in Papua New Guinea from the low-lying Carteret Island (an atoll of the autonomous region of Bougainville) to the mainland was considered to be the first case of climate change-induced migration. In the North, those displaced in 2005 by Hurricane Katrina in New Orleans were labelled by some as the first climate refugees in an industrialized country (Giroux 2006). The term "climate refugee" is used to describe a person who has decided (or was forced) to migrate in the face of climate change impacts. However, there is no official political definition and no official refugee status for affected populations.

The "climate refugee" or "climate change-induced migrant" has emerged as a key player in climate security discourses. In 2009, the Secretary General of the United Nations presented a report on *Climate Change and Its Possible Security Implications* (UNGA 2009) in which he recommended the development of a new legal status to protect those displaced by climate change. The following year, the international climate negotiations under the United Nations Framework Convention on Climate Change (UNFCCC) recognized "climate change-induced migration" in the *Adaptation Framework* as an issue that might receive funding via the adaptation funds (UNFCCC 2010). The year after that, in a presidential statement, the United Nations Security Council recognized climate change as a threat to the national security of low-lying small island states that face submergence (UNSC 2011).

Most articles on the issue of climate refugees ask one or more of several questions. How many people are likely to be affected (for a critique, see Jakobeit and Methmann 2012)? How should climate refugee status be defined? Who should be eligible for help? And who should bear the costs (Docherty and Giannini 2009; Biermann and Boas 2010)? In this chapter, I ask a different set of questions. I investigate the climate refugee as a discursive construct in a contested narrative landscape on climate change. I ask: which (climate) policies are enabled by the changing problematizations of climate refugees? I draw on Foucault's (2007) governmentality studies in order to distinguish between three different discourses that have been problematizing climate refugees or climate change-induced migrants in different ways. For each discourse, I assess the policies mobilized and offer a critique of the policy implications.

The empirical analysis for this chapter draws on the most cited documents on climate (and environmental) refugees, from the publication of the first document by the UN Environment Programme in 1985 (El-Hinnawi 1985) until the present. Moreover, fieldwork was conducted at the fifteenth Conference of the Parties (COP15) to the UNFCCC in Copenhagen in December 2009, where I carried out a series of interviews with activists, small island delegations, and scientists, and recorded all side events with a focus on climate change-induced displacement. All interviews and side events were transcribed. The collected material was then subjected to a discourse analysis, which sought to distinguish between competing ways of framing the issue of climate change-induced migration. I analysed which discourses were dominant or marginalized at which point in time and discuss the policy implications of each discourse. Moreover, I analysed the ways in which dominant discourses were contested and by whom.

In the remainder of this chapter I present the findings of that discourse analysis. I distinguish between three discourses of climate refugees and climate change-induced migration. Each discourse was dominant at some point in time, but all three still exist today. The second section introduces the discourse that constructs "millions of climate refugees" as a threat to states' national security. I associate this alarmist discourse with Foucault's concept of sovereign power and argue that it mobilizes defence against the presumed threat. The third section presents the discourse that considers climate change as a threat to the human security of people (mostly in developing countries). This discourse charges industrialized countries with "saving" climate refugees. I argue that it mobilizes risk management to secure people in line with Foucault's liberal biopower. Finally, the fourth section introduces the discourse that accepts dangerous levels of climate change as inevitable and recommends migration as a rational strategy of adaptation to rising sea levels. I show how advanced liberal government renders affected populations responsible for helping themselves. For each discourse, the resulting policies are discussed and the policy implications highlighted.

I conclude that all three discourses on climate refugees or climate change-induced migration contribute to legitimizing the displacement of millions of people. By presenting dangerous levels of climate change as inevitable and by making resilience the new leitmotif of climate policy, the third and most recent discourse depoliticizes the issue of climate change in a radical way. From a critical perspective, it is necessary to contest this "partition of the sensible" (Rancière 2004) in order to restore our ability to address the causes of climate change. I highlight the need for an alternative framing in which calls for emission reductions in industrialized countries can be legitimately made and compensation for damages suffered requested. By contesting the dominant discourses on climate change-induced migration, I seek to open up new ways of acting on climate change and to contribute to an ecological geopolitics.

Defending against the vulnerable

In the late 1980s and early 1990s, the environmental refugee emerged as a figure that the industrialized countries were told to fear. At the time, there was rising concern that ongoing environmental degradation could lead to mass displacement and possibly even violent conflict, especially in fragile states (Homer-Dixon 1999; see Meierding, Chapter 4, this volume, for further discussion). The emerging chaos in the Global South would lead to the North being overwhelmed by waves of refugees. This section presents evidence of such discourse with regards to "climate refugees" and analyses its policy implications.

In 2007, Greenpeace Germany was responsible for media headlines warning of more than 200 million so-called "climate refugees" worldwide by 2037. Greenpeace had commissioned an academic study on the issue of climate refugees (Jakobeit and Methmann 2012) and launched its findings at a big press conference. At a time when international climate negotiations were deadlocked, the environmentalists hoped to attract public support for emission reductions. Another example of this discourse is Michael P. Nash's documentary *Climate Refugees* (2009), which includes the phrase "the human face of climate change" in the subtitle. In the film, Nash interviews people who have been severely affected by changing weather patterns around the world. Shown in close-up, they describe losing their homes, their communities, and their children to extreme weather events, often with tears in their eyes. These images are linked together by a narrative of forced migration; all of these people will have to migrate sooner or later as a result of climate change – and, as the film suggests, the migration route will lead straight from Global South to North. At one point a globe appears on the screen with red arrows illustrating the presumed migration patterns of the affected populations. As all of these arrows terminate in the United States or Europe, a threat to national security is clearly established. Such migration will be a problem, the film tells us, because it will cause conflict, possibly violence, and even climate wars. Nash interviews several high-ranking members of the US military establishment who provide evidence for these claims. After one hour, footage of an exploding nuclear bomb is shown to illustrate the possibility of a third world war as a result of unmanaged mass migration. The film ends with a dramatic appeal to use energy-saving light bulbs and recycle household waste, because such behaviour will make a difference in the fight against global warming. No government-level political action is demanded; finding solutions to the problem is delegated to the individual consumer.

These two examples represent a discourse that constructs climate refugees as threats to national security in the Global North – a mass of people who should be feared. Moreover, migration is conceptualized as a problem that should be avoided – and where it cannot be avoided, security in the form of defence is advocated, for example by strengthening border installations. According to a Pentagon study, the results of abrupt climatic change will be a world in which "[d]isruption and conflict will be endemic features of life" (Schwartz and Randall 2003: 22). Other examples of this discourse include a report written by retired US military personnel (CNA Corporation 2007) and Harald Welzer's *Climate Wars* (2012). All estimates of the number of future climate refugees cited in the reports mentioned above can be traced back to a study by Myers and Kent (1995), which has since been discredited for its flawed methodology. While this discourse of national security is no longer dominant, it is still popular, especially in the United States, where the CNA Corporation recently published a follow-up to its original 2007 study (CNA Military Advisory Board 2014). (Simon Dalby analyses this document in detail in Chapter 6, this volume.)

Which forms of governing climate change in general and climate refugees in particular are incited by the national security discourse? The Copenhagen School (Waever 1995) has developed a theoretical framework in order to investigate the social construction of issues as security issues. They argue that the successful "securitization" of an issue like climate change could elevate it to the level of high politics and enable politicians to circumvent democratic procedures and/or adopt extraordinary measures. The drastic articulation of climate change as an existential threat to the survival of nation states could justify such a political state of exception if it were to be accepted by a "relevant" audience (Waever 1995). Extraordinary measures like "the Security Council adopting resolutions to impose emission targets, and even military measures against polluting factories" (Trombetta 2008: 599) might then be adopted in such a case. While climate change has actually made it onto the agenda of the UN Security Council three times (2007, 2011, and 2013), not a single resolution on the subject has yet been passed. The reason for this is that the majority of developing countries contest the framing of climate change as a security issue and insist, instead, that it is an issue of sustainable development to be negotiated under the UNFCCC. The climate negotiations have continued without taking much notice of the climate security discourse and without any breakthroughs since the adoption of the Kyoto Protocol in 1997.

We are also not witnessing the militarization of climate politics at international climate negotiations (Oels 2013). Instead, there has been a climatization of security policy (Oels 2013): that is, the consideration of climate change impacts in military planning and training (Dalby 2014). Following Foucault's governmentality lectures, national security discourse can be seen to mobilize a sovereign economy of power – namely, one based on the use of force as a last resort. Globalization has led to decentralized production networks, many of which draw directly or indirectly on resources and agricultural products from conflict-prone regions. Defence is the favoured strategy of securing global economic flows where mass migration and resulting violent conflict threaten to disrupt these global production chains. Hartmann (2010) has argued that the climate security discourse may, in the end, legitimize military interventions in destabilized post-disaster regions. Moreover, it contributes to the long-standing securitization of migration. As Bigo (2007) has highlighted in his work on migration, governments draw on ever more sophisticated strategies of surveillance (such as satellite surveillance) in order to identify presumably "dangerous" individuals within the masses that migrate. The climate security discourse may lead to more investment in border technologies to keep out climate migrants.

In conclusion, this discourse spreads fear about climate refugees. It constructs climate refugees as a threat to the national security of states, and mobilizes defence as the mode of securing. However, there is no evidence that this discourse either facilitates emission reductions or leads to the militarization of climate policy. Instead, it seems to have led to the climatization of defence policy and the securitization of migration policy.

The discourse that urges us to fear climate refugees can be criticized on a number of counts. First, it is undeniably racist. For instance, Andrew Baldwin (2013) has highlighted the racist stereotypes reproduced in the film *Climate Refugees*. The documentary establishes Western experts as superior and developing country victims as in need of help, thereby echoing colonial stereotypes of "the dangerous South" (Dalby 1996, 2009). Climate refugees are presented as both threat and victim simultaneously (Baldwin 2013: 1479).

Second, raising fears about "millions of climate refugees" helps to legitimize restrictive migration policies and the militarization of borders. The UN Commissioner for Refugees, Antonio Guterres, in an official side event to the UN Climate Conference in Copenhagen in 2009, spoke strongly against linking migration and security:

> I don't think migration is a security threat . . . Now, migration is a security threat for those countries that believe the way forward is to close their borders and not let anybody in. Of course, the best way to justify that is to invoke security reasons, like if every migrant was a potential terrorist . . . And traffickers and smugglers . . . they develop because there is not enough opportunity for legal migration.

Roland Emmerich's movie *The Day after Tomorrow* (2004) offers a powerful reminder that anybody can become a refugee. The movie focuses on a temperature drop – as does the aforementioned Pentagon study (Schwartz and Randall 2003) – that forces US citizens to flee to warmer countries, such as Mexico. The images of desperate Americans risking their lives in order to overcome their own sophisticated border defences are, of course, highly ironic.

Third, while many proponents of this discourse claim otherwise, it does not facilitate emission reductions. Some studies do not even mention the causes of climate change. For instance, Schwartz and Randall's (2003) report for the Pentagon was written under the George W. Bush administration, which denied that climate change was human-made. As a result, there is no talk of causation in the study, and the evidence presented in it is divided. Moreover, if the causes of climate change are mentioned at all, the problem of climate change is individualized. As mentioned above, in the concluding section of *Climate Refugees*, the documentary focuses on behaviour that everybody should adopt to limit climate change. However, there is no talk of the need for new policies, such as climate laws or international treaties.

I conclude that this discourse can be criticized for being racist and for potentially making migration more difficult.

Protecting vulnerable populations (interventionism)

During the 1990s, a new way of framing climate change and climate refugees emerged. In this new discourse, the international community of states was said to be responsible for "saving" climate refugees from rising sea levels, enabling a regime of liberal biopower. The discursive shift was facilitated by the rise of the concept of human security, which was advocated by the United Nations Development Agency in its *New Dimensions of Human Security* report (UNDP 1994). Moreover, after the end of the Cold War, the international community launched a number of so-called "humanitarian" military interventions in conflicts, overriding national sovereignty in the name of universal human rights, thereby ushering in an interventionist era (Chandler 2012).

This discourse, which was dominant from the 1990s to the early 2000s, conceives of climate change as a risk to humans that is calculable and therefore manageable. Global climate modelling provided the technology for calculating emission pathways that were presumably "safe" for the majority of people. However, there was also an acknowledgement that current weather variations were already dangerous for many people, the biosphere, and the economy. Climate scientists were increasingly able to map populations who were considered to be particularly vulnerable to climate change, including those whose livelihoods might be destroyed and who might therefore be forced to migrate. Governments were charged with targeting interventions on the most vulnerable (Methmann and Oels 2014) – the "dangerous" groups, as Foucault would say. Such vulnerable populations were considered to be in need of assistance by the international community.

This interventionist stance on the issue of climate change-induced migration was facilitated by the rise of the human security discourse and the spread of humanitarian interventions in the 1990s. In his report on *Climate Change and Its Possible Security Implications*, the Secretary General of the UN establishes climate change as a threat to human vulnerability; however, he refrains from explicitly using the politically contested term "human security" (UNGA 2009). Instead, he states: "Adequately planning for and managing environmentally induced migration will be critical" (UNGA 2009: 17).

The overall policy rationale of the human security discourse is risk management, based on scientific calculations and cost–benefit analysis (Oels 2013). Rather than banning greenhouse gas (GHG) emissions in general, risk management seeks to keep them at a presumably "safe" level. The two-degree target, which seeks to keep the average global temperature increase to two degrees Celsius above pre-industrial levels, is a good example of this. Global warming is not stopped, but the rate of warming is supposed to be manageable for nature, the economy, and humanity. According to Foucault's governmentality lectures, we can classify this as a regime of liberal biopower (Dean 2010). Liberal biopower uses statistics and scientific calculations to manage the well-being of the population at large. In risk management, particular attention is paid to those who deviate statistically from the norm(al), and interventions are targeted at them.

In the case of climate change, this implies that policy must focus on those who are most vulnerable to climate change: for example, those living in so-called "climate hot spots." According to the human security discourse, governments should identify these groups and target their interventions on them. Where national governments fail to protect their own populations – for example, in so-called "fragile" states of the Global South – Northern governments may legitimize themselves to intervene by humanitarian or military means (Hartmann 2010). The human security discourse creates "the 'humans' requiring securing" (Duffield and Waddel 2006: 2) so that they can then be legitimately subjected to Northern interventionism.

In *Climate Change and Its Possible Security Implications*, the Secretary General of the UN demanded new legal frameworks to protect those displaced by the impacts of climate change, especially those from low-lying small island states rendered stateless as a result of rising sea levels (UNGA 2009). The academic literature has discussed various options for a legal framework, ranging from a recognition of "environmental persecution" under the UN Refugee Convention (Conisbee and Simms 2003: 33), via a protocol to the UNFCCC (Biermann and Boas 2010), to a stand-alone convention (Docherty and Giannini 2009). However, there seems to be no political support for a refugee status of any sort. The issue was taken up at the international climate negotiations in 2010 in Cancun under the new label of "climate change-induced migration." This wording was chosen to clarify that no asylum will be offered to climate refugees and no claims can be made against the industrialized countries. Instead, Article 14f of the *Adaptation Framework* agreed at Cancun lists climate change-induced migration as eligible for funding from adaptation funds (UNFCCC 2010). However, even this had not been operationalized at the time of writing (late 2014).

The human security discourse constructs populations in climate hot spots as helpless victims in need of Northern assistance. Their political agency is denied, and others are empowered to speak on the refugees' behalf. Claudia Aradau (2004) has demonstrated that this sort of "politics of pity" is highly problematic. The discourse on human security can be criticized for reducing citizens to humans, thereby denying people's political agency. As Agamben (1998) has highlighted, the refugee is the citizen's other; she or he is constituted by lacking much of what defines the citizen.

Second, the creation of a new category of climate refugee perpetuates a flawed refugee regime. In an interview conducted in Copenhagen in 2009, two No Borders activists argued that introducing climate refugee status would merely serve to "divide us as human beings" and "redefine who you exclude." It would build on a refugee regime in which

> the only way you can possibly hope to stay in the UK is if you are a really good victim. People have got to feel really sorry for you to be allowed to stay. You've got to have, like, persecution, scars, you know, like all this evidence.

Instead, No Borders works towards "freedom of movement for all," which would allow people to say, "'You know, I just wanted to move.'"

Third, populations under threat of displacement by climate change strongly resist their classification as "climate refugees." In interviews with ambassadors from low-lying small island states, McNamara and Gibson (2009) repeatedly heard, "We do not want to leave our land." Similarly, in my own interviews with NGO representatives from small island states, they all stressed that they do not want to migrate. However, if they are forced to leave, they wish to travel as labour migrants (which is currently highly restricted), not as designated refugees:

> Our president made it clear – and we totally back him up on this – that we don't want to be called climate refugees, because we are not. As I keep saying, we are so tiny, but we are very proud people and we would never dream to be a burden to other countries. I would never dream to come to one country and be called a climate refugee. I'd love to move with dignity and respect, with merit. And that's why we back our government in trying to get skills ... [so that] we move in with the skills and contribute to the country, rather than being a burden on the country.

Carol Farbotko insists that small island people in the Pacific have political agency; they are not passive, helpless victims, as the Western media claims. More importantly, they have a long tradition as seafarers upon which any self-determined migration could build (Farbotko 2012; Farbotko and Lazrus 2012).

I conclude that the human security discourse seeks to prepare the ground for Northern interventionism. However, those threatened by climate change do not want to be rescued; they do not want to leave their land.

Inciting resilience in vulnerable populations (post-interventionism)

The idea that populations affected by climate change have political agency and can take their own precautions has informed the latest discourse on what is now officially called "climate change-induced migration." This discursive shift took place in 2011 when the UK's Government Office for Science published *Migration and Global Environmental Change*. Very much in line with neoliberal ideas of self-optimization and empowerment, this report reframed climate change-induced migration as a rational strategy of adaptation to climate change impacts.

This latest shift in the discourse is based on the assumption that dangerous levels of climate change cannot be prevented. Of course, this has become more likely since the climate negotiations failed spectacularly in Copenhagen in 2009. Indeed, current climate variability is already dangerous for many people around the world, and, with little or no progress on drafting a new comprehensive climate treaty, an average global warming of more than three degrees Celsius is now considered likely (Rogelj et al. 2010). Under such conditions, tipping points in the global climate system might be reached, and these could lead to the collapse of the Gulf Stream and the death of the Amazon rainforest (Lenton et al. 2008). The global climate system is theorized as non-linear, so any changes are highly unpredictable and radically contingent, but the consequences remain potentially catastrophic: "With such warming, there is little uncertainty over whether extreme impacts will occur, only when they will happen, and to what extent they will affect specific locales" (Mabey et al. 2011: 43). As a result, "[t]he threat of climate change is high-impact and high-probability" (Mabey et al. 2011: 84).

The latest discourse has reframed climate change as "environmental terror" (Duffield 2011: 763) that strikes unpredictably. The idea that science could define "safe" emission pathways has been dropped. Boykoff et al. (2010: 53) argue that, due to unknown climate sensitivity and unknown carbon cycle dynamics, it is almost impossible to identify a "safe" level of atmospheric carbon dioxide. In the face of the unknown, new sources of knowledge – such as scenario planning studies and worst-case scenarios – are spreading and forming the basis for a politics of preparedness.

In a world presumably facing environmental terror, practices of securing are informed by the concepts of resilience and preparedness. The founder of the resilience concept, C.S. Holling (1973: 14), defines resilience as a social or ecological system's ability to "absorb change and disturbance and still maintain the same relationships between populations or state variables." Resilience can range from maintenance via adaptation to transformation. The resilient subject is "conceived only as an active agent, capable of achieving self-transformation" (Chandler 2012: 217). Resilience is in line with what Foucauldians would call advanced liberal government (Dean 2010; Oels 2005) and what Evans and Reid (2013: 11–12) have called the "neoliberalized care of the self." In a regime of advanced liberal government, the individual is rendered responsible for self-optimization by building a strong social network for emergency situations. In a radically contingent world, the capacity for adaptive emergence, for reflexive self-transformation, is key for survival (Dillon 2007). Resilience mobilizes the vulnerable towards programmes of self-help, to foster "their entrepreneurial abilities and technical skills" (World Bank 2010: 130–1).

The Intergovernmental Panel on Climate Change (IPCC) published its *Fifth Assessment Report* in 2013/14 (for further discussion, see O'Lear, Chapter 7, this volume). In the concluding chapter, Working Group II, tasked with focusing on socio-economic aspects of climate change impacts, acknowledges that:

> climate change [is] a threat to sustainable development ... as a result, transformational changes are very likely to be required for climate resilient pathways – both transformational adaptations and transformations of social processes that make such transformational adaptations feasible.
>
> *(Field et al. 2014: 1106)*

In the latest scientific publications, climate change-induced migration is reconceptualized as a rational strategy of coping with climate change (Black et al. 2011). *Migration and Global Environmental Change* argues that migration is an excellent strategy of what it calls "'transformational' adaptation to environmental change ... [which] in many cases will be an extremely effective way to build long-term resilience" (Government Office for Science 2011: 7). The migration of some can allow those staying behind to engage in adaptation and survive, if enough remittances are sent (Scheffran et al. 2012). However, not all people in regions exposed to climatic changes and extreme weather events can become resilient. Some might become even more vulnerable in the process of migration. Many might be unable to migrate and end up trapped in dangerous places. Policy actors emphasize that the risk of what they problematize as "maladaptation" remains high. Therefore, careful planning is required in such cases (Asian Development Bank 2012: 47), focusing interventions on the "dangerous" or high-risk groups.

I conclude that the resilience discourse is rendering the potential victims of climate change responsible for taking their own precautions. It does so by drawing on advanced liberal technologies that seek to push individuals towards self-transformation, including relocation.

As critics of the concept of resilience have pointed out, resilience is not the same as sustainable development. The former is about survival; it defines a bottom line of "sheer survivability" (Evans and Reid 2013: 9). There is no promise of political rights, no talk of human rights, no minimum standards of existence. As a result, resilience is much less than adaptation, mitigation, and sustainable development. This enables a new form of government based on post-interventionism; the Global North is no longer responsible for fixing the situation in the South (Chandler 2012). Of course, this might lead to decreasing flows of assistance from North to South.

The most contested aspect of the resilience discourse is its assumption of the inevitability of disaster that many people around the world will presumably have to prepare for and then endure. At stake in this debate are alternative visions of geopolitical futures (McNamara and Gibson 2009). It is simply unacceptable that affected populations are to suffer from extreme weather events on a regular basis or even disappear from the map. At the international climate negotiations in Warsaw (COP19) in November 2013, Yeb Sano, the spokesperson for the Filipino delegation, made a strong statement on this subject:

> [W]e refuse as a nation to accept a future where super typhoons like Haiyan become a fact of life. We refuse to accept that running away from storms, evacuating our families, suffering the devastation and misery, having to count our dead, become a way of life. We simply refuse to ... We can stop this madness. Right now.
>
> *(Sano 2014)*

Sano pledged to fast during the conference until significant progress on emission reductions was made. (This was also an expression of solidarity with many of his fellow Filipinos, who had no food for three days after Typhoon Haiyan hit the islands.) His main point was that climate change can still be mitigated; that it is still possible to act and prevent many of the potentially catastrophic climate change impacts. It might be difficult to persuade people in the industrialized countries to abandon their SUVs, but that does not mean that it is impossible. The No Borders activists in Copenhagen in December 2009 suggested that those displaced by climate change should adopt the slogan "We are here because you drive SUVs" to establish that Northern lifestyles are a principal cause of their migration.

It is only by contesting the inevitability of climate change that demands for reductions in GHG emissions can be legitimately raised. As one small island state delegate declared during a side event at Copenhagen:

> I am not going to turn and run away from my home because of the rising of the sea. No! ... My biggest problem is people are not listening when we say, "Cut your emissions." They don't listen. They don't care. That's my problem. Not the water.

The chairperson of the Alliance of Small Island States insisted in an interview: "We are still hoping that all the members of the international community, especially in the General Assembly, will take necessary action in time to address climate change in such a way that it won't affect us" (quoted in McNamara and Gibson 2009: 480). However, the resilience discourse rejects the possibility of a geopolitical future in which climate change does not have a devastating impact.

Conclusions: restoring the political in climate change politics

This chapter is based on the assumption that climate refugees should be investigated as a discursive construction that serves some purpose, to paraphrase

Robert Cox (1981). However, looked at more closely, it is clear that there is more than one climate refugee discourse. I have shown that climate refugees were first constructed as a threat to be feared; then as people who need our support; and, finally, as people who are capable of self-help and self-determined relocation. Over time, even the label for people displaced by climate change has changed from "climate refugees" to "climate change-induced migrants." The former was dropped in order to signal that no claims to asylum could be made. This might be considered progress, as I have shown that the affected populations have no desire for refugee status. In fact, most of them flatly refuse to view themselves in such a light. If they are eventually forced to leave their homelands, they would prefer to do so as labour migrants rather than refugees.

However, there are some good reasons to be sceptical about the resilience discourse. It depoliticizes the issue of climate change, making it seem as if nothing can be done about it, as if it is merely a "fact of life" (Sano 2013). The displacement of millions of people is no longer framed as a political scandal, but as a rational strategy of adaptation. The loss of millions of livelihoods is reframed as unfortunate, but inevitable (McNamara and Gibson 2009). At the same time, the citizens of the industrialized nations continue to drive their SUVs without a trace of guilt. Populations under threat of displacement contest the inevitability of climate change as something for which they must prepare. They highlight that it could still be slowed and mitigated and its impacts minimized. They refuse to accept the inevitability of the disappearance of their homes from the map. They refuse to accept the need to live "dangerously" (Evans and Reid 2013). They insist that, if the large emitters act in time, their homes can still be saved. In fact, many climate change impacts *could still be* avoided, but they *will not be* if the industrial nations continue to show no interest in changing their behaviour. The resilience discourse effectively obscures this potential for action and naturalizes the impacts of climate change.

The resilience discourse most notably fails to problematize the causes of climate change: the fossil fuel-based capitalist system of production and consumption, which produces most of the GHG emissions. As Swyngedouw (2010: 223) has highlighted, if the fossil fuel-based capitalist system is problematized in climate security discourse at all, it is only in talk of an "aberration" of an otherwise flawless system that can be fixed. In fact, the same elites who created the problem in the first place are now charged with solving it (Swyngedouw 2010: 223), using market measures. Swyngedouw mentions markets for trading carbon and carbon offsets as examples of attempts to solve climate change by using the very methods that caused it (see also Glover's discussion, Chapter 2, this volume). Ecological geopolitics make it very clear that the decisions that we make about how we produce and consume in the Anthropocene are shaping the planet's geology for future generations. Emissions are deeply embedded in the lifestyles of the industrialized countries. A questioning of the very patterns of fossil fuel-based capitalist production and consumption is required if we are to address the problem of climate change at its roots and pre-empt the need for mass relocation.

References

Agamben, G. (1998) *Sovereign Power and Bare Life,* trans. D. Heller-Roazen, Stanford, CA: Stanford University Press.
Aradau, C. (2004) "The perverse politics of four-letter words: Risk and pity in the securitisation of human trafficking," *Millennium: Journal of International Studies,* 33(2): 251–79.
Asian Development Bank (2012) *Addressing Climate Change and Migration in Asia and the Pacific,* Manila: ADB.
Baldwin, A. (2013) "Racialisation and the figure of the climate change migrant," *Environment and Planning A,* 45(6): 1474–90.
Barnett, J. (2001) *The Meaning of Environmental Security: Ecological Politics and Policy in the New Security Era,* New York: Zed Books.
Biermann, F. and Boas, I. (2010) "Preparing for a warmer world: Towards a global governance system to protect climate refugees," *Global Environmental Politics,* 10(1): 60–88.
Bigo, D. (2007) "Detention of foreigners, states of exception, and the social practices of control of the banopticon," in P.K. Rajaram and C. Grundy-Warr (eds) *Borderscapes: Hidden Geographies and Politics at Territory's Edge,* Minneapolis, MN and London: University of Minnesota Press, pp. 3–33.
Black, R., Bennett, S.R.G., Thomas, S.M., and Beddington, J.R. (2011) "Climate change: Migration as adaptation," *Nature,* 478: 447–9.
Boykoff, M.T., Frame, D., and Randalls, S. (2010) "Discursive stability meets climate instability: A critical exploration of the concept of 'climate stabilization' in contemporary climate policy," *Global Environmental Change,* 20: 53–64.
Chandler, D. (2012) "Resilience and human security: The post-interventionist paradigm," *Security Dialogue,* 43(3): 213–29.
CNA Corporation (2007) *National Security and the Threat of Climate Change,* Alexandria, VA: CNA Corporation.
CNA Military Advisory Board (2014) *National Security and the Accelerating Risks of Climate Change,* Alexandria, VA: CNA Corporation.
Conisbee, M. and Simms, A. (2003) *Environmental Refugees: The Case for Recognition,* London: New Economics Foundation.
Cox, R.W. (1981) "Social forces, states and world orders: Beyond international relations theory," *Millennium: Journal of International Studies,* 10(2): 126–55.
Dalby, S. (1996) "The environment as geopolitical threat: Reading Robert Kaplan's 'Coming Anarchy,'" *Ecumene,* 3(4): 472–96.
Dalby, S. (2009) *Security and Environmental Change,* Cambridge: Polity.
Dalby, S. (2014) "Environmental geopolitics in the twenty first century," *Alternatives: Global, Local, Political,* 39(1): 1–14.
Dean, M. (2010) *Governmentality: Power and Rule in Modern Society,* second edition, London: Sage.
Dillon, M. (2007) "Governing through contingency: The security of biopolitical governance," *Political Geography,* 26(1): 41–7.
Docherty, B. and Giannini, T. (2009) "Confronting a rising tide: A proposal for a convention on climate change refugees," *Harvard Environmental Law Review,* 33: 349–403.
Duffield, M. (2011) "Total war as environmental terror: Linking liberalism, resilience, and the bunker," *South Atlantic Quarterly,* 110(3): 757–69.
Duffield, M. and Waddell, N. (2006) "Securing humans in a dangerous world," *International Politics,* 43(1): 1–23.

El-Hinnawi, E. (1985) *Environmental Refugees*, Nairobi: UNEP.

Emmerich, R. (director) (2004) *The Day After Tomorrow*, Lionsgate.

Evans, B. and Reid, J. (2013) "Dangerously exposed: The life and death of the resilient subject," *Resilience*, 1(1): 1–16.

Farbotko, C. (2012) "Skillful seafarers, oceanic drifters or climate refugees? Pacific people, news value and the climate refugee crisis," in T. Threadgold, B. Gross, and K. Moore (eds) *Migrations and the Media*, New York: Peter Lang, pp. 119–42.

Farbotko, C. and Lazrus, H. (2012) "The first climate refugees? Contesting global narratives of climate change in Tuvalu," *Global Environmental Change*, 22: 382–90.

Field, C.B., Barros, V.R., Dokken, D.J., Mach, K.J., Mastrandrea, M.D., Bilir, T.E., Chatterjee, M., Ebi, K.L., Estrada, Y.O., Genova, R.C., Girma, B., Kissel, E.S., Levy, A.N., MacCracken, S., Mastrandrea, P.R., and White, L.L. (eds) (2014) *Climate Change 2014: Impacts, Adaptation, and Vulnerability: Part A: Global and Sectoral Aspects: Contribution of Working Group II to the Fifth Assessment Report of the Intergovernmental Panel on Climate Change*, Cambridge: Cambridge University Press.

Foucault, M. (2007) *Security, Territory, Population: Lectures at the Collège de France 1977–78*, New York: Picador.

Giroux, H.A. (2006) "Reading Hurricane Katrina: Race, class, and the biopolitics of disposability," *College Literature*, 33(3): 171–96, http://findarticles.com/p/articles/mi_qa3709/is_200607/ai_n16717316/print, accessed 29 January 2013.

Government Office for Science (2011) *Migration and Global Environmental Change*, London: Government Office for Science.

Hartmann, B. (2010) "Rethinking climate refugees and climate conflict: Rhetoric, reality and the politics of policy discourse," *Journal of International Development*, 22(2): 233–46.

Holling, C.S. (1973) "Resilience and stability of ecological systems," *Annual Review of Ecology and Systematics*, 4: 1–23.

Homer-Dixon, T. (1999) *Environment, Scarcity and Violence*, Princeton, NJ: Princeton University Press.

Jakobeit, C. and Methmann, C. (2012) "'Climate refugees' as a dawning catastrophe? A critique of the dominant quest for numbers," in J. Scheffran, P.M. Link, and J. Schilling (eds) *Climate Change, Human Security and Violent Conflict: Challenges for Societal Stability*, Berlin and Heidelberg: Springer, pp. 301–14.

Lenton, T.M., Held, H., Kriegler, E., Hall, J.W., Lucht, W., Rahmstorf, S., and Schellnhuber, H.J. (2008) "Tipping elements in the earth's climate system," *Proceedings of the National Academy of Sciences*, 105(6): 1786–93.

Mabey, N., Gulledge, J., Finel, B., and Silverthorne, K. (2011) *Degrees of Risk: Defining a Risk Management Framework for Climate Security*, London: Third Generation Environmentalism.

McNamara, K.E. and Gibson, C. (2009) "'We do not want to leave our land': Pacific ambassadors at the United Nations resist the category of 'climate refugees,'" *Geoforum*, 40(3): 475–83.

Methmann, C. and Oels, A. (2014) "Vulnerability," in C. Death (ed.) *Critical Environmental Politics*, London: Routledge, pp. 277–86.

Myers, N. and Kent, J. (1995) *Environmental Exodus: An Emergent Crisis in the Global Arena*, Washington, DC: Climate Institute.

Nash, M.P. (director) (2009) *Climate Refugees*, Trulight Pictures.

Oels, A. (2005) "Rendering climate change governable: From biopower to advanced liberal government?," *Journal of Environmental Policy and Planning*, 7(3): 185–208.

Oels, A. (2013) "Rendering climate change governable by risk: From probability to contingency," *Geoforum*, 45: 17–29.

Rancière, J. (2004) "Who is the subject of the Rights of Man?," *South Atlantic Quarterly*, 103(2–3): 297–310.

Rogelj, J., Nabel, J., Chen, C., Hare, W., Markmann, K., Meinshausen, M., Schaeffer, M., Macey, K., and Höhne, N. (2010) "Copenhagen Accord pledges are paltry," *Nature*, 464: 1126–28.

Sano, Y. (2013) "Typhoon Haiyan: We cannot afford to procrastinate on climate action," *Guardian*, 11 November, www.theguardian.com/world/2013/nov/11/typhoon-haiyan-philippines-climate-change, accessed 27 June 2014.

Scheffran, J., Marmer, E., and Snow, P. (2012) "Migration as a contribution to resilience and innovation in climate adaptation: Social networks and co-development in Northwest Africa," *Applied Geography*, 33: 119–27.

Schwartz, P. and Randall, D. (2003) *An Abrupt Climate Change Scenario and Its Implications for United States National Security*, Washington, DC: Environmental Media Services.

Swyngedouw, E. (2010) "Apocalypse forever? Post-political populism and the spectre of climate change," *Theory, Culture & Society*, 27(2–3): 213–32.

Trombetta, M.J. (2008) "Environmental security and climate change: Analysing the discourse," *Cambridge Review of International Affairs*, 21(4): 585–602.

United Nations Development Agency (UNDP) (1994) *New Dimensions of Human Security*, New York: UNDP.

United Nations Framework Convention on Climate Change (UNFCCC) (2010) "Cancun agreements," http://unfccc.int/resource/docs/2010/cop16/eng/07a01.pdf#page=2, accessed 26 October 2012.

United Nations General Assembly (UNGA) (2009) *Climate Change and Its Possible Security Implications*, report of the Secretary General, A/64/350, New York: UNGA.

United Nations Security Council (UNSC) (2011) "6587th meeting, 20 July 2011, 3 p.m., S/PV.6587 (Resumption 1)," New York: UNSC.

Waever, O. (1995) "Securitization and desecuritization," in R. Lipschutz (ed.) *On Security*, New York: Columbia University Press, pp. 46–86.

Welzer, H. (2012) *Climate Wars: What People Will Be Killed for in the 21st Century*, London: Polity.

World Bank (2010) *Development and Climate Change: World Development Report 2010*, Washington, DC: World Bank.

13
TOWARDS ECOLOGICAL GEOPOLITICS

Climate change reframed

Simon Dalby and Shannon O'Lear

In the introduction to this volume we posed the question of how to reframe climate change – how to think differently about the issue, and above all how to think about it as much more than an old-fashioned "environmental" issue. Obviously, climate is about environment, but the point is that it impinges on so many parts of human life that it has to be much more comprehensively addressed than has been the case with previous "environmental" crises. Climate is part of a widespread transformation of the biosphere that humanity has set in motion. Climate change is about the really big questions of how the future will be shaped not only in terms of "the environment," but also in terms of a range of interconnected systems of economics, agriculture, transportation, finance, and society, all intertwined with dynamics of power and influence. These questions of earth governance, and who will decide the future configurations of the biosphere, suggest that global ecology is – whether politicians and pundits like it or not – increasingly shaping geopolitics (Hommel and Murphy 2013).

The planet can no longer be taken for granted as the given stage for great power rivalries, struggles for imperial dominance, or international prestige. These things persist, but increasingly they are played out in circumstances that are changing precisely because of how geopolitics is practised. Now, the questions that matter are more about how geopolitics will shape the future climate than about how climate changes geopolitics (Dalby 2014). The huge transformations that the biosphere is undergoing now require us to think of ecology as part of geopolitics, and of geopolitical practices as active components that shape the new geological epoch we are creating – the Anthropocene. We obviously need to think much more carefully about how to link governance, technology, and environment to deal with what Victor Galaz (2014) calls the "Anthropocene gap" between changing ecological realities and human institutions. Many of the chapters in this volume have done just that. The final question is how their efforts might be pulled together to suggest both new academic questions and provocative political insights that lead us to act in new ways that have the potential to make a practical difference.

Grappling with the implications of our current circumstances requires us all to rethink our place in the biosphere and to reframe the kinds of questions that we ask as social scientists, and choose to act on collectively as citizens. As Porter and Hulme (2013) note, it requires thinking about framing matters in terms of the specific issue of climate change in new ways rather than falling back on "generic frames" of environment, progress, innovation, and so on in our thinking – which, as James Smith and Shaun Ruysenaar (Chapter 9) remind us, the biofuels debate in particular does so often. Quite deliberately, none of the chapters in this volume deals in much detail with the traditional framing of climate change in terms of mitigation, adaptation, and geoengineering – the standard "MAG" policy frames.

There is much more to climate communication than merely matters of frames (Cox 2010), to be sure. Nevertheless, thinking about which frames are currently used – and which other frames might be and to what effect – matters greatly because the framings that are used in political discourse shape how we all think and act, and crucially how we might think and act differently. We tend to focus on what lies inside a frame. Indeed, the purpose of a frame is to showcase a particular view or perspective. Throughout this book, however, the contributing authors have demonstrated the importance of paying attention to what lies outside a particular frame. Although Naomi Klein (2014) might be overstating the situation by suggesting that climate "changes everything," her points about the sheer scale of transformations, the failures of conventional political actions to reduce greenhouse gas (GHG) emissions – never mind reverse them – and the need for urgent action to make fossil fuel production and consumption socially unacceptable emphasize how dramatically the climate issue needs to be reinterpreted. This book has sought to demonstrate how current framings of climate change often fall short of providing the kinds of perspective, insight, and pathways forward that are needed at this time.

Leigh Glover (Chapter 2) puts the case that, so far at least, climate has not generated a reframing that steps outside the given modern assumptions of industrial society. Climate discussions usually still assume the inevitability of continued economic growth as the solution to all societal ills. Now, of course, growth is reframed as "sustainable development," in contrast to earlier versions of economic growth that we have learned are unsustainable. Yet, the growth assumption remains intact. While there are compelling reasons why people in poorer parts of the world will strive to increase their wealth to improve health, nutrition, education, and life chances, in general the sustainable development frame still asserts continued economic growth for all, including those who have made the situation unsustainable in large part by their massive use of fossil fuels. Sustainable development is about managing the current situation, about using technology to innovate our way out of the climate change predicament, and frequently about using market-based governance mechanisms to direct policy. It does not offer a helpful reframing that departs from harmful practices.

Two decades ago, the geopolitical framework for addressing climate change was established in the formulation of the UNFCCC. That institution has served as a significant frame for interpreting climate change. As several chapters of this book have shown, the inadequacy of that institutional structure as a mode of ecological geopolitics is now abundantly clear. The Kyoto Protocol and the system of emphasizing common but differentiated responsibilities have allowed economically developed states to act sluggishly, if they have bothered to act at all, on the significant aspects of climate change. Simultaneously, that structure of governance has postponed poorer states' efforts to reconsider their development strategies to avoid the fossil fuel-based modes of economy that have caused climate change. All of this suggests that the political economy of fossil fuel-based growth at the heart of modernity is the problem that has to be tackled quite directly if we are to develop an ecological geopolitics that offers governance arrangements to challenge the perpetuation of fossil fuel-powered economic expansion in an already rapidly changing world. The dangers of not rethinking these basic assumptions of modernity are profound.

One key attempt to challenge the taken-for-granted assumptions in modernity is to ask whether climate change is dangerous. After all, the UNFCCC is premised on avoiding "dangerous" anthropogenic interference in climate systems. However, how much interference is truly dangerous, and how we might know this, is not so simple, as Chris Russill points out when he examines the problems of detection and specifying danger (Chapter 3). The way in which the IPCC was established focused on consensus, incremental change, and the metric of global mean surface temperature. Nothing in this formulation was likely to generate alarm or specifications of immediate danger. In the 1990s, climate appeared to be changing slowly, with long-term rather than immediate consequences. In Russill's terms, the detection idiom was a mode of communication ill-suited to expressions of danger. Clearly, science has detected an unambiguous human cause in current climate change, but that is producing neither much-needed policy actions nor larger cultural changes that would be appropriate in the circumstances.

In the latter stages of Russill's chapter, he suggests that notions of tipping points and ecological changes that are anything but gradual, incremental temperature changes have been gaining attention in recent years. The Anthropocene formulation and concerns about extreme events offer ways of framing matters that more obviously allow expressions of alarm and warnings about coming dangers. They do so because they understand the situation in terms of complex systems and potentially dangerous tipping points of humanity as parts of a complex biospherical transformation. Climate change is about much more than carbon dioxide; appropriate political action and policy innovation have to demand much more than economic analyses of possible risks from changing levels of carbon dioxide to an otherwise supposedly stable system. In this way, Russill's chapter reflects an earlier concern voiced by Erik Swyngedouw (2010) regarding the perception or construction of an external-to-humans "nature" that could be stabilized through the proper management steps. Swyngedouw expands on Žižek's argument that "Nature does not exist!" (Žižek 1992: 38, quoted in Swyngedouw 2010: 301), at least not in the way that we have constructed it as a universal backdrop to human existence. Instead, there are multiple possible socio–nature relations. Invoking "nature" (or "the environment" or even "climate change," broadly speaking) in such unquestioned, universal terms forecloses more thoughtful enquiry into the dynamics of power enabled by these empty signifiers. Indeed, drawing distinct boundaries can be detrimental.

The conventional framing of climate change (i.e., idealizing economic growth, accepting the promise of sustainable development, maintaining a focus on carbon dioxide as the most important measure) can provide a false sense of stability "here" and draw our attention to other forms of instability "there." The concept of environmental security persists in part because political instabilities related to climate change are interpreted as threatening "our" stability and requiring a military response. Emily Meierding's methodological proposal to disconnect climate change from conflict (Chapter 4) emphasizes the importance of the implicit geographical framing in much of the discussion about climate change. Assuming, as much scholarship on climate and conflict does, that scarcities in the poorer parts of the planet are the problem, and focusing ever more sophisticated social science techniques on tracing links between rural disruptions in the Global South and local political conflicts, directs attention away from the sources of climate change in metropolitan consumption, much of it in the affluent Global North. This selective framing of climate and conflict serves to reduce the North's responsibility to deal with the problems of fossil fuel consumption, both directly, in terms of fuel used, and indirectly, in terms of the larger economic dislocations caused by this mode of economy. Nevertheless, Meierding also notes, crucially, that most of the studies that try to find statistical correlations between environmental factors and political conflict fail to do so.

Meierding's chapter suggests that there is a much wider range of human responses to climate change than simply fighting over scarce resources. By focusing attention on violence, rather than many other non-violent, cooperative, and sometimes plainly altruistic responses, social scientists often miss important possibilities. They do so by perpetuating assumptions of conflictual social relations and using statistical methods that present what appear to be clear-cut findings in media reports that popularize their results. Such scholarship and media coverage feed into an overall geopolitical framing of "them and us" and of fear on the part of "us" obscuring the causes of climate change in the fossil fuel-powered global economy by discussing putative rural resource shortages instead.

This theme of division also comes through clearly in Paul Routledge's discussion of the political ecology of Bangladesh and the invocation of justice in the face of dispossession and vulnerability (Chapter 5). He considers discourses of climate justice by working "from the ground up" in Bangladesh, a state that is highly vulnerable to storms in the immediate future, and to rising sea levels that are part of the environmental transformation caused by the warming planetary climate over the longer term. Crucial to his analysis is the key political ecology insight that people's access to land and other resources is shaped by the structure of rural agrarian economics. Frequently, the rural poor are absent from decision-making about adaptation to climate change and their voices not heard in the expert deliberations of NGOs, governments, bureaucrats, and development planners. As Routledge emphasizes, the poor and landless people who are often most vulnerable are in an antagonistic relationship with the formal economy and its financial mechanisms. They are often forced into direct political action to gain access to land and always face the danger of dispossession by commercial enterprises eager to enlarge their holdings. However, this tension is not a matter of climate change causing political violence; it is primarily a question of violent social injustice, which may now be aggravated by storms and rising seas.

Routledge's chapter also makes abundantly clear that assumptions that there is some consensus about how to adapt to climate change are missing crucial parts of the politics in rural areas in many parts of the world. The "we are all in this together" assumption in the "post-political" formulations of climate policy serve to obscure the power relations that matter in how rural ecologies are being transformed, and in how people are becoming increasingly vulnerable to both environmental and economic hazards. Thinking about ecological geopolitics requires that these calls to justice, and the practical politics of thinking about how to resist the expansion of the capitalist social relations that have caused much of contemporary climate change, can be understood and confronted in trying to build more just and sustainable societies.

While Meierding and Routledge look at local campaigns in the rural Global South, Simon Dalby (Chapter 6) starts with the view from Washington, DC, and the US military's attempt to use a conflict framing of climate as a way to generate attention for the issue in the United States. He examines attempts by some senior military officers and the CNA think-tank to frame climate change as a matter of US national security. Formulating climate change as a potential threat multiplier, and a conflict catalyst, suggests imminent social disruptions that will affect US interests on a sufficiently large scale to warrant treating them as national security issues. Securitizing climate and raising it to a high-level potential threat has not been successful in generating widespread agreement internationally on dealing with climate in terms of security or, more recently, in persuading many politicians in Washington to agree to fund climate security activities on the part of the military – all of which suggests that framing climate as a threat requiring the apparatus of national security may be an ill-considered approach to the issue. Many of the discussions that point to militarization as an appropriate policy response to climate change overlook one fundamental question: what is being secured and for whom (Dalby 2009)? Indeed, applying familiar, state-centric notions of security to the spatially complex and multidimensional issue of climate change represents a failure of securitization in the current context.

This failure of securitization suggests that climate change is perhaps better understood as a matter of sustainable development. Looking to economic investments, the possibilities of innovations in energy systems, and intelligent infrastructure planning may be more useful, because they focus on the future and routine matters rather than exceptional measures. It is a mistake to frame climate change in terms of potential dangers rather than in terms of the possibilities of creatively altering the direction in development policy. An ecological geopolitics should build for the future rather than fear social disruptions and prepare to deal with them violently. But, as Glover makes clear in Chapter 2, sustainability needs to be divorced from contemporary growth assumptions that assume that technical fixes and market management will somehow change economic systems enough to avert dangerous disruptions in coming decades.

In Chapter 7, Shannon O'Lear investigates the pernicious effects of conventional territorial assumptions when measuring the causes of climate change and assigning responsibility. While it might appear to be sensible to start with territorial states because they are the basic units of governance in the contemporary world, counting emissions in terms of what happens within the territories of particular states omits a crucial part of the story. Because of international trade, items made in one part of the world are frequently used somewhere else. The current practice of assigning responsibility for GHG emissions to where things are made rather than to where they are used shifts responsibility away from the consumers who use the products to the producers. Effective attribution of responsibility for the embodied carbon in commodities requires that the counting methods reflect the final use of the product, but the current systems do not do that. Fixing this geographical sleight of hand would lead to much better allocation of the responsibilities for coping with climate change.

One of the most obvious points about geopolitics and ecology is tackled directly here. While it is generally understood that environmental processes do not respect political boundaries, the issue of embodied carbon adds another important twist to our reliance on territorial strategies to respond to climate change. Territorial jurisdictions may allow certain forms of compliance while simultaneously generating injustice. Accounting for embodied carbon in a way that holds end-use consumers responsible for their behaviour recognizes the mobility of carbon and its entanglement in complex economic systems of exchange. Rigid territorial strategies that do not allow cross-boundary flows to be governed sustain an unworkable frame locked on to the geopolitical map. Geography matters in questions of justice, and technical issues of how carbon use is counted have important political consequences. Focusing on the end use of carbon, much of it in the rich countries of the North, rather than repeatedly blaming China (where much contemporary manufacturing takes place) for climate change, would much more fairly attribute responsibility and increase the political pressure on those who have caused most of the climate problem to start to fix things by cutting their carbon consumption. This holds true not merely for individual consumers, but more significantly for businesses, industries, and trade arrangements that continue to widen the gap between those who benefit from resource use and those who are left with negative externalities not captured in market prices. Framing ecological matters in the convenient arrangement of territorial jurisdictions obscures key issues of the politics of measurement. This is more than a matter of *what* is counted. *Where* things are counted is crucial in ecological geopolitics.

One of the things that is counted most frequently in climate change discussions is the rapidly rising level of carbon dioxide in the global atmosphere. This measurement, which was fluctuating around 400 ppmv at the time of writing (2014), is a key part of the earth system science that has been monitoring global developments over the last few decades. Reducing global environmental change to some key metrics – and carbon dioxide is the most high-profile variable, as Russill argues in Chapter 3 – encourages a view of the world as a single system susceptible to being managed in various ways. Given the alarm about the increasingly severe consequences of climate change and the very slow pace of international efforts to address the problem, serious discussions have now started about how to intervene in the earth system and "manage" the amount of solar radiation reaching the surface of the planet. That we are even discussing solar radiation management is an indication of how rapidly climate change is moving in the wrong direction and, as such, in many ways this is an act of desperation. However, as Thilo Wiertz makes clear in Chapter 8, there is always a risk that some profound political questions will get lost amid the technical details in discussions of geoengineering. The focus on technical solutions also implies that these are feasible and – despite a long history of technical and social problems with major engineering projects, – practically controllable. Who decides how hot the planet will get is likely to be a very contentious discussion indeed!

Key for any serious discussion of ecological geopolitics is this question of how technical matters substitute for politics and how the language of science, measurement, and lumping emissions from all sources into single measurements obscure responsibility for them and preclude social responses that are sensitive to particular contexts. The second Copernican revolution, the shift to understanding life as a key part of the earth system that has shaped it profoundly for hundreds of millions of years, is a necessary starting point for ecological geopolitics. However, reducing this understanding of life to matters of simple measurement, or opaque computer models that end up suggesting technical manipulation of the atmosphere, eviscerates the politics from the discussion by focusing on technical matters rather than human values. It does so by replacing collective human deliberation with decisions imposed by an engineering elite, and in the process runs the risk of perpetuating the fossil fuel-powered transformation of the biosphere that caused the problem in the first place by supposedly providing a "get out of jail free card" for contemporary modes of consumption. On the other hand, geoengineering might yet be "needed" if political and economic decisions fail to tackle GHG emissions seriously and soon. Such dilemmas will be at the heart of ecological geopolitics in coming decades.

Smith and Ruysenaar (Chapter 9) look to another policy option that has been widely discussed in terms of reducing fossil fuel consumption: the possibilities of literally growing fuel by converting crops into biofuels. In theory, at least, this approach to fuel generation would seem to be carbon neutral. The fuel generated by biofuels might appear to be renewable because carbon drawn from the atmosphere by plant photosynthesis is a key part of the vegetable matter that is used to make the liquid fuel. However, such biofuels are components of complicated agricultural economies that use substantial amounts of fossil fuel for fertilizer, farm machinery, transport, and processing. Biofuel production areas are more than just another way to use carbon fuel sources. These lands and production systems are parts of local ecologies and food supply arrangements, too. Framing biofuel as a commercial opportunity perpetuates existing economic institutions and extends the reach of their practices instead of questioning the kinds of ecologies they are creating. It is that kind of rethinking about human-created ecologies that climate change seems to require urgently.

Framing biofuels in their larger economic and agricultural contexts suggests that a narrow focus on carbon alone may lead to perverse consequences. Substituting biofuels for fossil fuels deals only with fuels, not with the larger economy that is transforming so many parts of the biosphere. Biofuel production may also replace food production, and that possibility raises the question of whether vehicles are more important than people in the modern world. That question, additionally, has profound political consequences in terms of climate justice and the priorities for dealing with ecological transformation. In terms of ecological geopolitics, biofuels threaten to perpetuate a framing of the world that looks to rural landscapes as resource sources and does so in ways that overlook the subsistence needs of the people who live there as well as the nutritional needs of people elsewhere in the global economy. Once again, uncritical discussions about fossil fuels focus attention on industrial sources of continued consumption rather than more carefully rethinking how to organize urban life in ways that do not require ever larger resource extractions from rural hinterlands.

Rethinking our reliance on these processes and practices will require reframing the issues in many ways. Andrew Szasz (Chapter 10) clearly suggests that this reframing will have to be done by organizations that have not normally been viewed as environmentalists. Crucially, given the United States' political importance as well as its historically large contribution to GHG emissions, these reframings will have to happen in US politics in addition to elsewhere. Given that a fundamental point of the denial industry is to reject the relevance of scientific findings, Szasz emphasizes that simply challenging the denial industry by generating ever more scientific findings is unlikely to generate the political impetus to develop the necessary policy innovations. Appeals for climate action in terms of environment have not yet generated any effective political action. The kind of rethinking that is needed will have to engage mainstream America in new ways, and this is starting to happen; US military organizations, the insurance industry, and "mainline" churches are framing climate as an issue that challenges their core interests.

If these parts of the larger US polity engage productively with the American public, Szasz suggests that a "meta-frame" of climate change as an existential threat to global human society might emerge with sufficient political impetus to overcome the denialist frame that has stymied so many climate change policy efforts in the United States. This meta-frame is a key component for ecological geopolitics. It involves a shift of perspective from one in which humanity is on the earth to one where humanity is understood as a force that is actively shaping the future of the earth. It also requires recognizing that environmental change is both caused by human action and simultaneously requires human action to cope with that change. Humanity is an active player, not a passive spectator, in the climate change drama. Ecological geopolitics is all about that meta-frame and the consequences of taking it seriously as the basis for political action in an interconnected world.

Most models of climate change suggest that extreme weather events are likely to become more frequent, more severe, or probably both in coming decades. Planning to cope with these events is now prudent public policy, but how government bodies frame what needs to be insured against and prepared for is not as simple as it might at first appear. While Szasz focuses on the long-term, big picture of the insurance industry – and the threat posed to it by climate in the United States – Kevin Grove (Chapter 11) focuses on the often hidden politics around climate adaptation and the financialization of risks relating to hurricanes and other natural disasters. In looking carefully at the Caribbean Catastrophic Risk Insurance Facility (CCRIF), he shows that the politics behind how dangers are framed, and who and what will be protected in the event of a disaster, is not a straightforward matter of helping people in distress. It is much more a technocratic exercise that links financial markets, risk assessments, and the institutions of states, rather than primarily a concern for the well-being of people who are potentially in harm's way.

The focus on a biopolitical analysis suggests, crucially, that people are frequently caught up with institutions that are ill-equipped to deal with rapid environmental change and the hazards that climate change will bring. Adaptation is frequently considered in narrow ways by governments, and the CCRIF is much more interested in ensuring the continuity of government than in protecting or helping vulnerable citizens with few response options. If government institutions are not designed to facilitate adaptation and deal with the injustices that make people vulnerable, then there is an important political point here for any notion of ecological geopolitics. Reconceptualizing governance requires thinking beyond states and international financial arrangements to consider how power works to shape people's preparations for a changing world, and how ecosystems and human systems may need to be reconfigured quite dramatically to deal with poverty in ways that facilitate the necessary flexibility to cope with both gradual environmental changes and very rapid ones in the form of disasters. If resilience allows systems only to return to their precarious and unjust state prior to a disaster, then such policies and social arrangements are parts of the problem that needs to be addressed, rather than an appropriate set of responses that will lead to a more sustainable future.

Angela Oels' discussion of "climate refugees" (Chapter 12) complements Grove's analysis by showing how various forms of security have been invoked when dealing with adaptation to climate, but have frequently been rejected as the most appropriate policy framing. In part, refugees and migrants have been portrayed as threats whenever they cross borders. Hence, they are doubly victimized: first, they are framed as displaced persons stripped of rights, citizenship, and political agency; and then they are portrayed as a threat to societies that might be disrupted by their arrival. Not surprisingly, this framing of climate in terms of national security has been rejected by many states that prefer to think in terms of sustainable development strategies and assistance with various forms of adaptation that might make their societies less vulnerable to the disruptions that are undoubtedly coming.

Refusing to accept a framing of climate refugees as a threat, and insisting that people should not have to live as victims and that they should not be considered as inevitable casualties of unavoidable transformations, highlights the politics of climate change that is so frequently silenced in technocratic and administrative framings of the issue. Here, Oels emphasizes the need to think differently about climate change by focusing on the source of the problem, not simply on managing some of the unfortunate consequences. She points to key political questions that are frequently forgotten or glossed over in the various policy frames of both national and international negotiations – and, indeed, in social science formulations and academic investigations, too. Her chapter makes it impossible to avoid the key question: who is climate change policy for?

This framing, along with those of Routledge and Grove, in particular, emphasizes a key point about formulating ecological geopolitics. It demands that the view from the margins – from displaced peoples, subsistence farmers, and vulnerable dwellers in informal settlements – forms a key part of the discussion. Simply reworking modernity, as Glover suggests, is not enough to think seriously about reconstructing societies that can deal with the consequences of fossil fuel-powered "development." Szasz's meta-frame of an existential threat to global human society emphasizes that climate change has the potential to disrupt globalized interconnected human societies to the extent that reframing development as a matter requiring alternatives to fossil fuel-powered economies – not a matter of yet more coal- and petroleum-powered economic "growth" – is now key to any notion of a sustainable future for human civilization.

Much of the concern about climate change over the last decade has been framed in terms of the vulnerability of human lives and artefacts to increasingly severe weather events and rising sea levels. In North America, this discussion has crystallized around discussions of Hurricane Katrina, which led to the flooding of New Orleans in 2005, and Superstorm Sandy, which caused flooding in New Jersey and New York in 2012. Emblematic of the latter event is the inundation of the New Jersey boardwalk and fairgrounds, and especially the image of the partially submerged roller-coaster frame that we use on the cover of this book. We chose that image not just because of the visual pun on the theme of unworkable frames but because it highlights the dangers of building in vulnerable places and the lack of foresight that is frequently employed in human endeavours.

Climate change requires rethinking many things, including, crucially, the assumption that past environmental conditions are a useful template for future plans. While climate change promises an increasingly bumpy ride for humanity, it is highly unlikely that the ride will be entertaining for those on the climate roller-coaster. Clearly, too, we will have to abandon some of the frames that we have used previously when planning buildings and entertainments in our new, increasingly artificial circumstances, where, among other things, rising sea levels are changing coastal geographies. Much of the content of this book has been about understanding the limitations of how we have framed climate change up to now, and how those frames constrain how we study contemporary transformations. Much more thinking about possible new frames is needed. That is beyond the scope of this volume, but it must not be beyond the scope of current scholars, policy-makers, and political activists. Escaping the constraints of many of the existing frames will, we hope, help humanity to deal with the tasks ahead, allowing more careful strategic thinking to shape the next phase of the Anthropocene so that societies can flourish in new ways and new times. In this quest, the authors in these pages are in good company; other scholars and commentators are making similar points elsewhere.

A recent book titled *Framespotting: Changing How You Look at Things Changes How You See Them* (Matthews and Matthews 2014) offers a subtly radical stance on environmental politics. First, the authors argue that it is important to recognize that we tend to perceive events, experiences, and discourses within certain framings that constrain our understanding. Second, they encourage us – once we are aware of these frames – to consider the ideas or connections that they may prevent us from seeing. As O'Lear emphasizes in Chapter 7, one of these framings is that the world tends to be understood in terms of states. The *Framespotting* analysis also argues that government-generated policies on climate change are widely thought to be synonymous with responses to climate change. These two points have been recurrent in this volume through the myriad ways in which contributing authors have questioned authority and investigated the shortcomings of an uncritical acceptance of current governance, finance, and techno-scientific structures. Reclaiming politics outside of these constraints is a way forward to developing creative and collaborative new framings for climate change (Connelly 2015; O'Lear forthcoming).

Different framings of climate change, such as mitigation and adaptation, involve dramatically different politics in how the problem and policy responses are conceptualized. This rethinking or reframing takes different forms depending on context, spatial scope, and even academic discipline. Political scientists, for instance, should be well suited to generating better public policy through an understanding of how incentives can influence politics and the importance of people's beliefs and expectations. It is surely possible to rethink frames of climate change in politically productive and "incentive compatible ways" (Keohane 2015: 25) involving fee and dividend schemes or other arrangements to gain public support for new ways of behaving. Keohane's argument has also been employed by the leading climate scientist James Hansen (2009), as well as the *Framespotting* authors, to suggest that simple fees on fossil fuel production coupled with payments to citizens to compensate for rising prices could quickly change the world's energy use patterns.

Accepting that we now live in the Anthropocene means acknowledging human capacity to alter the very conditions of life on this planet. Such an understanding should, reasonably, lead to the question: "what sort of climate do we want" (Caseldine forthcoming)? It also involves, as has been argued throughout this volume, thinking about relationships of power in different ways. Rather than starting with the familiar state system and then asking questions about how that system should respond to climate change, a different way forward is to focus on degraded planetary systems and related social, political, and economic implications, then try to identify appropriate forms and spaces of governance for their rehabilitation. An examination of the nitrogen cycle, for instance, and the significant impacts of applications of fertilizers and eutrophication of water bodies might lead us to focus on the idea of food sovereignty as a paradigm to guide us away from the agro-industrial model and towards small-scale agriculture, low-input farming, poly-cultures, and other approaches that balance power more immediately with human–environment interactions (Schroeder 2014). This directly connects to considerations of biodiversity that can lead us to consider the loss of indigenous cultures and knowledge, and the value of creating forms of governance that will foster and protect socially just property rights and forms of compensation.

What is more, our efforts to respond in meaningful ways to climate change will have to be "clumsy," in that we will likely need to return to them for repeated assessment and adjustment (Thompson 2013). There will need to be multiple, simultaneous, and contextually relevant responses to decelerate and redirect climate change processes: "silver buckshot" rather than a "single silver bullet" in the form of "a portfolio of approaches that would move us in the right direction, even though we cannot predict which specific ones might stimulate the necessary fundamental change" (Prins and Rayner 2007: 974). This is a very different approach from the conventional focus on the United Nations system and its plans to negotiate comprehensive binding agreements among the major states to solve the climate problem (Harris 2013). Indeed, it will be important to recognize different forms of power linked with climate and other environmental issues and the spaces that are connected, created, and degraded through these processes (O'Lear 2010). Shifting focus towards an ecological geopolitics means taking a different starting point and framing questions about planetary health, local well-being, and governance structures that recognize observable realities of climate change, inequities, and the profound injustices of the status quo.

This is not about abolishing the state system, but rather about reforming political systems by finding new modes of governance that steward societies, businesses, and patterns of consumption towards more just outcomes while reducing disruptions to natural processes. It means challenging unchecked power that serves to entrench societies and ecosystems further into unsustainable circumstances. The chapters in this volume provide some thoughtful examples of ways in which we may start to reframe climate change and reclaim politics. They all demonstrate that ecological geopolitics requires that these two processes – the reframing and reclaiming – go hand in hand. Imagining a more just and sustainable future for humanity requires both; acts of ecological citizenship rely on effective reinterpretations of our context and on rethinking how we must act now, in our particular contexts, to shape the future.

References

Caseldine, C. (forthcoming) "So what sort of climate do we want? Thoughts on how to decide what is 'natural' climate," *Geographical Journal*.
Connelly, J. (2015) "Book review: *Framespotting: Changing How You Look at Things Changes How You See Them*," *Global Policy*, 13 January, www.globalpolicyjournal.com/blog/13/01/2015/book-review-framespotting-changing-how-you-look-things-changes-how-you-see-them, accessed 30 January 2015.
Cox, J.R. (2010) "Beyond frames: Recovering the strategic in climate communication," *Environmental Communication*, 4(1): 122–33.
Dalby, S. (2009) *Security and Environmental Change*, Cambridge: Polity.
Dalby, S. (2014) "Environmental geopolitics in the twenty first century," *Alternatives: Local, Global, Political*, 39(1): 3–16.
Galaz, V. (2014) *Global Environmental Governance, Technology and Politics: The Anthropocene Gap*, Cheltenham: Edward Elgar.
Hansen, J. (2009) *Storms of My Grandchildren*, New York: Bloomsbury.
Harris, P.G. (2013) *What's Wrong with Climate Politics and How to Fix It*, Cambridge: Polity.
Hommel, D. and Murphy, A.B. (2013) "Rethinking geopolitics in an era of climate change," *GeoJournal*, 78: 507–24.
Keohane, R.O. (2015) "The global politics of climate change: Challenge for political science," *PS: Political Science and Politics*, 48(1): 19–26.
Klein, N. (2014) *This Changes Everything: Capitalism vs. the Climate*, Toronto: Knopf.
Matthews, L. and Matthews, A. (2014) *Framespotting: Changing How You Look at Things Changes How You See Them*, Alresford: Iff Books.
O'Lear, S. (2010) *Environmental Politics: Scale and Power*, Cambridge: Cambridge University Press.
O'Lear, S. (forthcoming) "Climate science and slow violence: A view from political geography and STS on mobilizing technoscientific ontologies of climate change," *Political Geography*.
Porter, K.E and Hulme, M. (2013) "The emergence of the geoengineering debate in the UK print media: A frame analysis," *Geographical Journal*, 179(4): 342–55.
Prins, G. and Rayner, S. (2007) "Time to ditch Kyoto," *Nature*, 449(7165): 973–5.
Schroeder, H. (2014) "Governing access and allocation in the Anthropocene," *Global Environmental Change*, 26: A1–A3.

Swyngedouw, E. (2010) "Trouble with nature: 'Ecology as the new opium for the masses,'" in J. Hillier and P. Healy (eds) *The Ashgate Companion to Planning Theory: Conceptual Challenges for Spatial Planning*, Burlington, VT: Ashgate, pp. 299–318.

Thompson, M. (2013) "Clumsy solutions to environmental change," in L. Sygna, K. O'Brien, and J. Wolf (eds) *A Changing Environment for Human Security: Transformative Approaches to Research Policy and Action*, London: Routledge, pp. 424–32.

Žižek, S. (1992) *Looking Awry: An Introduction to Jacques Lacan through Popular Culture*, Cambridge, MA: MIT Press.

INDEX

Aaht Sangathan 75
accumulation by dispossession 67, 69
acid rain 41–2
actor networks 143
actuarial statistics 176
adaptation 17, 23, 40, 71, 78, 92, 161–3, 171–5, 184–5, 188–9, 194–7, 204, 211–13
adaptive capacity 173, 184
adaptive strategies 60
advanced liberal government 189, 196
agricultural commodity markets 134
agricultural subsidies 134
agriculture: biotechnologies in 70; and productivity implications for conflict 58–9
agro-fuels 133; *see also* biofuels
An Inconvenient Truth (film) 31–2, 35
Anglosphere 92
Annex I countries 21, 106, 109
Annex B countries 106–7
antagonism 68–9, 71–3, 75
Anthropocene 3–5, 44–7, 95–7, 120–1, 128, 203, 205, 213–14; crisis 118
anthropogenic global warming 31, 119, 123
Arab Spring 91
armed conflict 52, 53–4, 56, 60, 63
Armed Conflict Location and Event Dataset (ACLED) 56
Asia 68, 72, 75–6, 90, 105, 137, 152, 178
Asia Peasants Coalition 75
assemblage 132, 133, 146, 174
asylum 195, 199; *see also* refugee

atmospheric carbon dioxide levels 23, 37, 119, 196, 209
atmospheric science 122

Bahamas 179
Bakken Oil Field, ND 1
Bangladesh 11, 68–78
Bangladesh Krishok Federation (BKF) 68, 71–8
Bangladesh Kishani Sabha (BKS) 68, 71–8
Beck, U. 36, 176
Beddington, J. 132
biodiesel 133
bioenergy 133, 144
bioethanol 133
biofuels 11, 132–9, 140–6, 204, 210; and developing countries 133–5, 142; emissions 135–8, 145
biopolitics 171–3, 174, 175–7, 211; political imaginary 172
biopower 172, 181, 190, 193–4
biotechnology in agriculture 70
borders 77, 106, 108–9, 111, 192, 195, 198, 212
boundary setting 136
burning embers 35
Bush administration 85, 193

calculation 3, 9, 42, 107, 110–11, 124–8, 145, 175, 176, 182–3, 194, 206, 208
capitalism 8, 25, 36, 48, 71–2, 77, 103, 110, 126; and carbon 18, 25–7, 156; neoliberal 70, 111; pro-growth 69

218 Index

carbon 1, 4, 7, 19, 38–42, 93, 100, 111, 196, 205, 210
Carbon Dioxide Assessment Committee 42
carbon emissions 7, 18, 37, 47, 71–2, 75, 94, 105–7; implications of 42–3
carbon footprint 1, 106, 108, 164
carbon neutrality 135, 138, 210
carbon offset 1, 9, 25–7, 103, 123, 199
carbon pricing 9, 18, 19, 27–8, 69, 137–8, 208
carbon regulations 42, 44–6, 109, 206
carbon removal 2, 95, 116, 137
carbon trading 25–6, 69, 108
Caribbean 171–2, 175, 178–84
Caribbean Basin Initiative (CBI) 179
Caribbean Canada Trade Agreement (CARIBCAN) 179
Caribbean Catastrophic Risk Insurance Facility (CCRIF) 171–3, 176–7, 180–4, 211
Castree, N. 121, 127–8
catalyst of conflict 84, 90, 91, 96–7, 207
catastrophe 35, 156; insurance 11, 171, 181–4; model 155, 172, 175–7, 181–2, 184; risk 172–3, 176–7, 179, 183; *see also* extreme weather
Catholic Climate Covenant 157–8
causes of climate conflicts 52, 56, 59, 62–3
Central Intelligence Agency (CIA) 154, 161
China 2, 21, 58, 91, 105–6, 108, 208; agriculture and conflict in 58, 91
Christian stewardship of the environment 159, 160
Church: Roman Catholic 157–8, 163–4; United Methodist 157, 159, 163; "mainline" Protestant 157, 164, 210
civil war models 54
Clark, W.C. 41, 44–7, 48
Climate Change, Gender and Food Sovereignty Caravan 75–7
climate change-induced migration 47, 58, 89, 90, 91, 152, 188–90, 193–4, 196–7; *see also* refugee
climate conflict models 54, 60–1
Climate Justice Action 68
Climate Justice Now! 68, 69
Climate Refugees (film) 190, 192, 193
climate sensitivity 37, 38, 40, 196
climate–society interactions 45; *see also* human–nature relations
climatization of security 95, 191, 192
cloud brightening 123; *see also* solar radiation management

CNA Corporation 11, 53, 58, 84–5, 86–7, 90–3, 94, 96, 152–3, 191, 207
coalitions against climate change 74–5, 152, 163
Cochabamba Declaration 69
Cold War 3, 7, 83, 101, 118–19, 179, 193
commodification 183
commodity chain 109, 111
common but differentiated 7, 21, 205
common pool resources 15, 25
commons 69, 71, 73–4; reproductive 74
complexity 18, 26, 36, 45, 102, 111, 121–2, 135–6, 138, 142, 144
complexity science approach 45
computer modelling 7, 37, 44, 119, 124, 209; simulation 125
Conference of the Parties (COP) 17, 68, 69, 75, 155, 189, 198
conflict 4, 8–9, 11, 27, 52–63, 67–8, 71, 83, 84, 87–91, 109, 121, 143, 152, 160, 166, 190–3, 206–7; *see also* security, war
conflict datasets 56–7, 59
conflict multiplier 8
Congress (US) 85, 93, 150, 153, 161, 163
Congressional committees: testimonies at 150, 161, 163
consumption-based: approaches 107–9; emissions 110; accounting 107–9
contact language 36–7, 39, 46, 48; definition of 37
Copenhagen Accord 17
coping capacities 126, 174, 197, 208
coping strategies 60–1, 70
corn as biofuel 133; *see also* maize
cost–benefit analyses 102, 194
Crutzen, P. 3, 44, 46, 47, 116, 119, 120

Daly, H. 24
dangerous interference 6, 14, 17, 22–3, 33, 35–8, 39, 205
Davies, O. 180–1
decision-making xi, 17, 71, 103, 104–5, 126, 127–8, 137, 141, 146, 174, 207
dematerialization 24
Democratic Party (US) 84, 150
denial 43, 94–6, 150–2, 164, 165–6, 210–11; *see also* sceptics
Department of Defense (US) 53, 91–2, 93, 153, 161, 191, 193
detection idiom 36–41, 46–8, 205
deterritorialization 70
developing countries 53, 59, 69, 107, 108, 133–4, 136, 139, 142, 144, 191–2; and poverty alleviation 89, 134; *see also*

less developed countries, non-Annex I countries
development (economic) 2, 18, 28, 62, 68, 78, 88, 89, 93, 96, 132–5, 143, 177–81, 205, 208, 212
disasters 11, 31, 35, 40, 48, 54–5, 57, 60, 62, 63, 71, 83, 85, 86–7, 92, 97, 152, 172, 173–4, 176–84, 192, 197, 211; *see also* catastrophe, Hurricane
discourse 3, 4, 5–6, 41, 138, 144–6, 177–8, 199, 213; policy 37, 47, 68, 189–90; political 83–4, 94–5, 96–7, 188, 191–3, 195–8, 204; public 11, 25, 33, 35–6, 38, 46–7, 48, 68, 77–8, 206; scientific 118, 151, 193–5
displacement 63, 70, 135, 152, 189–90, 195, 199
Doha Amendment 19
domestic stability (US) 154
Duffield, M. 194, 196

earth system science 47, 95, 97, 117–21, 124–6, 209
eco-efficiency 23
ecological debt 69
ecological geopolitics 4–5, 8–10, 28, 48, 111, 118, 190, 199, 203, 205, 207–12, 214–15
ecological system 18, 28, 44, 104, 111, 120, 196
ecological systems perspective 44
ecological well-being 5
economic growth 3, 17, 18, 20, 24, 58, 88, 106–7, 204, 206
economists: role in policy 24, 25, 40, 43, 44, 71
edible crops: as source of biofuel 133
efficiency 23–4, 28, 120, 135, 145
El Niño Southern Oscillation (ENSO) 123; as measure of climatological shift 57
embodied carbon 100, 106–11, 208
emissions 2, 7, 11, 14–22, 24, 25–8, 31, 37, 38–43, 47, 69, 71–2, 75, 88, 94, 101, 103, 105–11, 116, 117, 135–8, 142, 159, 194, 198, 199, 204, 208, 209, 210
emissions gap 19
emission measures 110
emission permits 27
emission reductions 14, 19, 107, 109, 190, 192, 198
emissions statistics 107–10
emissions trading 18, 19, 25, 26–7, 103
energy crisis 41
energy policy 41

energy security 38, 41
Energy Security Act 1980 41
environmental contradictions of capitalism 20, 26, 28
environmental governance 3, 9, 172, 174
environmental management 16, 28, 102, 118, 120, 124
environmental security 6, 54, 57, 60, 84, 181, 184, 206; studies of 54, 57, 60
environmental terror 196
environmentalisms of the poor 71, 73
ethics 121, 124, 125–6, 127
Evangelical(s) 157, 160, 163, 164; Environmental Network 160–1
experts 26, 36, 48, 91, 118, 143, 151, 153, 192
extreme weather 5, 36, 40–1, 44–7, 57, 76, 86, 90, 152, 153, 155, 156–7, 162, 166, 178, 190, 196–8, 205, 211; and insurance 155–7, 171–3, 175–7, 178–9, 181–5; *see also* Hurricane

Feasibility Study (South Africa) 137, 142–3
Fifth Assessment Report (IPCC) 18, 86, 90, 93, 101, 104, 105–6, 108–9, 197
financialization 28, 173, 181, 183–4, 211
financialized disaster management 172, 181–5
food security 71, 143, 145
food shortages 90, 153; *see also* resource conflict
food sovereignty 73–5, 77, 214
fossil fuel 2, 3, 8–10, 16, 23–4, 26, 36, 41, 69, 71, 87, 92–7, 103, 105, 119, 132, 136–7, 150, 199, 204–6, 209, 210, 212, 214; *see also* agro-fuel, biofuel
fossil fuel-based capitalism 23–4, 26, 69, 95, 96, 103, 199, 204–5, 206, 212
Foucault, M. 117, 172, 175, 177, 178, 189, 191, 193, 194
Fourth Assessment Report (IPCC) 18, 19, 85, 89
futures 46, 77, 119, 121, 123–4, 126, 176, 184, 198

Geneva Association, The 155
geoengineering 24, 116–18, 120–1, 123–4, 127–8, 204, 209
geophysics: in defining climate change 37–40, 44, 46–8
geopolitics 9–10, 11–12, 100–1, 103–4, 181, 203, 208
GHG (greenhouse gas) 2, 8, 17, 22, 27, 116–17, 122, 145, 159

Index

GHG emissions 7, 14, 15, 16, 18–22, 24, 26, 28, 88, 101, 105–6, 108, 111, 116, 159, 194, 198, 199, 204, 208, 209
GHG neutrality of biofuels 132, 135–7, 142, 210
GHG sink 106, 137
global data 62, 101, 102
global economy 2, 8, 10, 20, 25, 91, 95, 96, 154, 180, 206, 210
global emissions 14, 18, 19, 106, 107
global fossil fuel subsidy 26
global mean surface temperature (GMT) 33, 37–40, 42, 46–7
Global North 67, 69, 76, 206
global science 126
global security 96
Global South 7, 67, 69, 77, 190, 194, 206, 207
global thresholds 39
global trade 108
global view 120
global warfare 8
globalization 4, 9, 12, 72, 75, 77, 93, 94, 109, 111, 191
Gore, A. 31, 33, 41, 47–8, 85
governance xi, 3, 6, 8, 9, 11–12, 16, 25–6, 36, 74, 87, 88, 103, 110, 121, 133, 138, 154, 173, 174, 203–5, 213, 214–15; of global biofuels production 138–42, 146; hierarchical structures of 17–18, 128, 135, 171–2, 208, 211
governmentality 117–18, 120–4; and Foucault 117, 189, 191, 194
Gramsci, A. 73
green governmentality 118, 120
greenhouse gas, *see* GHG
Greenpeace 190
Grenada 180–1

Hansen, J. 33–6, 38–9, 47–8, 214
Haraway, D. 126
holistic 126
Hulme, M. xii, 3, 6, 25, 37–40, 43, 44–5, 101, 126–7, 204
human–nature relations 117, 118, 120, 121, 126–8; or human–environment relationships xi, 5, 23, 103–5, 111, 118, 214
human responses to climate change 61, 206
human security 53, 84–6, 88–90, 94, 189, 193–5
human well-being 104, 166
humanitarian interventions 181, 193

Hurricane: Ivan 180–1; Katrina 31, 35–6, 48, 188, 212; *see also* catastrophe, extreme weather, Superstorm Sandy
hydrometeorological disasters 54–5, 57, 62–3
hydrometeorological measures 55, 59

IGBP (International Geosphere Biosphere Program) 119–20
IIASA (International Institute for Applied Systems Analysis) 44–6, 119–20
immigration 69, 151; *see also* migration, refugee
India 21, 76–7, 83, 105; agriculture and conflict 58
Industrial Revolution 4, 15–16, 22, 23
industrialization 15–16, 24, 28, 111, 156
Inhofe, J. 6, 151, 165
insurance 11, 47, 151, 154–6, 161–3, 165, 171–3, 175–9, 181–5, 210, 211
insurance imaginary 177
intellectual property rights 70
international climate change negotiations (or climate negotiations) 68, 87, 116, 188, 190–1, 194, 196, 198; *see also* IPCC, Kyoto Protocol, UNFCCC
international security 86
international trade 103, 106–7, 111, 180, 208
interventionism 193, 194–8; *see also* post-interventionism
intra-state armed conflict 52, 54, 57, 63
IPCC (Intergovernmental Panel on Climate Change) 2, 15–20, 23, 27, 33–6, 38–40, 44, 46, 70, 85–6, 89–90, 93–4, 100–6, 109–12, 116–17, 119, 151, 197, 205

Jamaica 180–1
jatropha 133, 145
justice 5, 6, 87, 125, 159, 207, 208, 214; climate justice 8, 11, 67–9, 71–8, 107, 173–4, 206, 210–11

knowledge–power relations 119
Kyoto Protocol 14, 17, 18–19, 21–2, 27, 111, 191, 205

Lalor, D. 179
land grab 70
land law 72
land occupation 69, 72–3, 77
land use change 107, 137
La Via Campesina (LVC) 74–5, 77

less developed countries: conflict and climate change in 53, 61–3; rural development and biofuels 132, 140; *see also* developing countries, non–Annex I countries
liberal-democratic governance 17, 25–6
life cycle analysis (LCA) 136–8, 140
Lobo-Guerrero, L. 175–6, 177, 183–4
Lomé Convention 179

McKinley, D. 93–5
maize 133, 145
management 11, 16–18, 20–8, 33, 41, 43–9, 67, 70–1, 87, 102, 116–18, 120–4, 145, 162, 173–8, 181–5, 206, 208, 209
maps of grievance 76–7
marginalized communities 128
market(s) 2, 3, 9–10, 18–19, 25–8, 69, 70–2, 73, 78, 102, 103, 128, 133, 134–5, 154–7, 162, 172, 178–85, 199, 205, 208, 209, 211
market environmentalism 18
market failure 18, 25
market measures 199
Middle East 90–1, 152
migrant 69, 188–9, 192–5, 199, 212; economic 154
migration 89, 90, 188–93, 193–8; and climate conflict 58, 91, 152; *see also* refugee
military intervention 192, 193
military (US) 1, 7–8, 9, 11, 47, 48, 56, 84–7, 88, 90–4, 96, 101, 119, 151–3, 161, 165, 190, 191–2, 193–4, 206, 207, 210
Millennium Development Goals 88
minerals–energy complex 146
mitigation 16–17, 18–19, 52–3, 63, 71, 74, 88, 90, 101, 106–7, 108, 116, 117, 120, 127, 135–6, 138, 143, 144–5, 171, 174, 176, 184–5, 197–8, 199, 204, 213
model misspecification 54–5
modelling (models): of catastrophe 155, 172, 175–7, 181–2, 184; of climate 7, 23, 37, 44–6, 117–19, 121, 122–7; of climate conflict 11, 52–5, 58–63; of climate sensitivity 37, 40; *see also* computer simulations
modernity 10, 15, 16–18, 20–1, 28, 95, 102–3, 205, 212
Movimento dos Trabalhadores Rurais Sem Terra (MST) 73
multilateral governance 140

narratives 10, 12, 61, 101, 110, 122–6, 132–3, 134, 144, 146, 189, 190

nation state 20–2, 26, 27, 71–2, 110, 191
National Association of Evangelicals 164
National Center for Atmospheric Research (NCAR) 102
national identity 102
National Intelligence Council (NIC) 53, 152–3
National Oceanic and Atmospheric Administration (NOAA) 35, 102
National Science Foundation (NSF) 102
national security 11, 48, 83–6, 88–91, 93–5, 97, 103, 151–4, 161, 165, 188, 189–92, 207, 212
nationalization of science 102
nature 9, 17–18, 28, 70, 118, 124, 194, 206; and religion 157–60
nature–society relations 73–4; *see also* human–nature relations
Navy (US) 152, 153, 161, 165
neoliberal (neoliberalism) 3, 16, 18, 69, 71, 73, 75, 102–4, 111, 196
neoliberal capitalism 69, 111
neoliberal governance 16
neoliberal market 103
neoliberal state 102, 104
new ecology of rule 78
non-Annex I countries 19, 21, 106, 109
non-edible crops: as source of biofuel 133
non-OECD countries 105
Nordhaus, W. 45
North America 1, 36, 38, 108, 136, 156, 179, 181, 184, 212–13
nuclear winter 119

oceanographers 33, 37, 153
OECD (Organization for Economic Cooperation and Development) 108; and agricultural commodity markets 134
offsets 9, 27, 199
optimization 118, 120–1, 124–6
overconsumption in developed states 53, 62, 69
ozone depletion 9, 41, 83

Paris 2015, climate treaty 2
participatory governance 174
peatland destruction in South East Asia 137–8
perfect storm 132–3, 135
performance 26, 126, 140
Plan B 116, 117
policy xi, 4, 7–8, 11, 14, 16, 19–20, 27–8, 37–8, 40, 44–9, 71, 95, 103–6, 140–2, 150, 175, 205, 210–11
policy deliberations 38, 43, 45

policy framework 37, 40, 43–4, 46, 68, 92, 110–11, 161–3, 204, 212
policy implications of 97, 138, 189, 190
policy-making xi, 9, 11, 28, 33, 39, 41–2, 52, 61, 83–6, 89–91, 94, 100–1, 109, 143–4; evidence in 138, 142–3, 145–6; scientists informing 102, 109, 116, 125, 134, 137, 144
policy narratives 44, 84, 88, 110, 144, 192, 194, 198–9, 205, 207
political ideology 84, 93–4, 95
political imaginary 172
politicization 3, 23, 71
politics 3, 9, 15, 21–2, 27, 36, 47–8, 68, 69, 71–3, 75, 77–8, 83–6, 92–5, 97, 100–1, 110, 116–21, 124–8, 134, 141–7, 150–5, 157, 161, 164–6, 171–5, 177, 181, 184–5, 188, 190–1, 194–7, 198–9, 204–7, 208–15
Pope Benedict XVI 157–8
Pope Francis 158
post-interventionism 195–8
post-political, or post-politics 23, 71, 77, 109–10, 171–5, 185, 207
postmodern environmentalism 28
power xi, 4–6, 9, 11, 12, 25, 67–9, 71–5, 91–2, 96, 103–5, 109–10, 119, 121, 126–8, 139–40, 146, 172–7, 178, 184–5, 189, 203, 206–7, 211–15; *see also* speak truth to power
power relations 69, 105, 110, 118, 121, 126, 128, 173, 178, 207
precipitation, impact on conflict 54–8, 62
prediction 36, 39, 52, 92, 101, 121, 126, 128
PRIO/Uppsala conflict dataset 56
pro-growth capitalism 69
projection 92, 124–6
Protestant churches 157, 164–5

radiative forcing 38, 40, 47–8, 124; *see also* solar radiation management
Rancière, J. 171–2, 189–90
rationality 117–18
redistribution of climate risks 117
reductionism 126
refugee 11, 69, 153, 188–95, 198–9, 212; *see also* climate change-induced migration, migrant
regulation 19, 37, 41–2, 67, 103, 184; biopolitical 174–6
reinsurance 47, 155–6, 163, 172, 179, 181–3
Reinsurance Association of America 156
relocation 188, 197, 199
remittances 60, 197

Republican Party (US) 84, 86, 92, 93, 95, 96, 150, 160
resilience 11, 156, 162–3, 171–5, 184–5, 189–90, 195–9, 211–12; ecology of 40, 44–5, 47–8
resistance 18, 63, 74
resource conflict 4, 9, 27, 54, 57–9, 61, 67, 206: narrative in media 53
resource enclosures 67
resource scarcity 9–10, 53, 54, 60, 67, 75, 152–4, 174
resource wars 89
responsibilities for climate change 63
risk 3, 15–17, 20–2, 33, 36, 39–42, 44, 47–9, 61, 63, 70, 85, 90, 107, 117, 127, 133–7, 140, 155–7, 162–3, 165, 172, 175–84, 193, 197, 205–6, 209, 211
risk management 43, 44–9, 162, 189, 194
risk pooling 176, 177, 179, 184
Roman Catholic Church 157, 164

scenario 23, 38, 40, 41, 46, 85, 122–5, 141, 144, 153, 181, 196
sceptics 22, 36, 150–1
Schelling, T. 43–5
Schellnhuber, J. 12–21, 36, 47, 126
science and policy 16, 33, 44–5, 100–5, 134, 144, 205, 212
science fiction 117, 122, 128
scientific consensus 23, 33, 37–9, 41–4, 109
scientific knowledge 2, 5–6, 18, 23–4, 38, 70, 105, 117–18, 127–8, 135, 142–4, 197
second commitment period 19
securitization 83–4, 94, 97, 102, 192, 207–8
security studies 54, 57, 60, 63; *see also* conflict, war
self-help 197, 199
sequestration 137; *see also* greenhouse gas sink
small island states 89, 96, 172, 188–9, 194–5, 198
Social Conflict in Africa Dataset 56
social domination of nature 28, 69
Social Forum 75
social movements 68, 71–5, 77
social responses to climate change 52–3, 60–3, 68, 71, 109–10, 172, 206, 209, 212, 214
social science(s) 3, 5, 10, 39, 42–3, 101, 111–12, 127–8
society–environment relationship 73–4; *see also* human–nature relations
socio-ecological change 174
socio-environmental change 69–71

solar radiation management (SRM) 116–18, 121–6; *see also* geoengineering
solidarity 68–9, 71, 74–7, 159, 198
South Africa 71, 108, 137, 140–8; shack dwellers in 73
South Asia Peasants Coalition 75
Southern Baptist Convention 160–4
sovereignty 103, 173–4, 193
spatial distribution 117, 125
speak truth to power 100, 143
state security 9, 173, 178–81, 184
state system 22, 46, 102–4, 107, 111–12, 214–15
state-centric security 8, 207
status quo 3, 10, 11, 26, 33, 105, 109, 214
Stengers, I. 122
Stern, Sir N. 18, 25
stewardship, environmental 22, 159, 160
structural adjustment programmes 67, 70
sub-Saharan Africa and climate conflict 61
subsidy 26
subsistence farming 74
sugarcane 133
Summary for Policymakers (IPCC) 101, 116
Superstorm Sandy 212
sustainable development 16–17, 23–4, 26, 68, 89, 95–6, 121, 191, 197, 204, 208, 212
sustainable management 120
Swyngedouw, E. 23, 71, 77, 109, 171–2, 199, 205, 206
Syria 63, 87, 91
systems approach 44, 118–20
systems ecology 44–5

technical knowledge 118, 135, 144, 185, 197
technological optimism 134–5
technology 11, 17, 18, 23–4, 28, 69, 101–2, 116, 118, 122, 128, 134, 144, 181, 193, 203, 204
temperature: impact on conflict 54, 56–8
territorial state 100, 103, 106–7, 109–11, 208
territorial trap 111
threat minimizer 88
threat multiplier 58, 84–90, 96, 152, 207
threshold: dangerous 23, 39, 40
tipping point 35, 36, 44, 47, 57, 196, 205
trade agreement 103, 108, 111, 179
transformation 8–10, 197, 203–7

transnational corporations (TNCs) 67, 74, 133
Tropic of Chaos (Parenti) 70
tropics: and biofuel production 137
two-degree target, or two-degree-Celsius increase 7, 14, 17, 19, 22, 39–40, 116, 194, 196
two-stage least-squared (2SLS) models 58

uncertainty 17, 48, 134, 155, 196
UNFCCC (United Nations Framework Convention on Climate Change) 2, 6, 15, 17–19, 21–3, 26–7, 33, 40, 46, 106–7, 109–10, 155, 188–9, 191, 194, 205
United Nations *Human Development Report* 88, 193
United Nations Environment Programme (UNEP) 14–15
United Nations Secretary General (UNSG) report 84, 188
United Nations Security Council (UNSC) 53–4, 87
United States (US) 21, 44, 48, 83, 85–6, 91–7, 101–2, 106, 108, 119, 134–5, 150–4, 157, 163, 190–1, 211
US Global Change Research Program (USGCRP) 33
US National Intelligence Council 153

victims 11, 61, 192, 194–5, 197, 212
violence 53, 59–61, 70, 72, 87, 121, 175, 180, 190, 206–7; *see also* conflict, war
violent contention 52, 54, 59–60, 62, 63
virtual carbon 106, 108
virtual ethics 125
virtual technologies 121, 124
vulnerability 17, 47, 48, 67, 70, 85, 86, 100, 150, 173–5, 182, 193, 212

war 54, 57, 60, 83, 84, 86, 91, 94, 101; *see also* Cold War
war of position (Gramsci) 73
war on coal (Obama's alleged) 92
water shortages 54, 57, 153; *see also* resource conflict
Watts, M. 78
weather and conflict 57, 87, 152, 190
well-being 5, 28, 67, 88, 104, 166, 184, 194, 211
Westphalia: cancer of 22, 103
World Bank 7, 19, 52, 57, 165, 180–2, 197